D0540201

Natural gas and national policy

LEONARD WAVERMAN

Natural gas and national policy:

a linear programming model of North American natural gas flows

UNIVERSITY OF TORONTO PRESS

Toronto and Buffalo
Printed in Canada
ISBN 0-8020-1832-7
Microfiche ISBN 0-8020-0157-2
LC 70-185743

To the woman in my life

Contents

Tables

Figures

Preface

This book began as a doctoral dissertation written for the Department of Economics, Massachusetts Institute of Technology. The teaching of Professor Morris Adelman interested me in the energy sector. Without the help and encouragement of Professors Adelman, Frank Fisher, and Paul MacAvoy, the thesis would never have been completed. A fellowship grant from Resources for the Future Inc. allowed me to devote full time to developing the ideas. Both the National Energy Board and the Federal Power Commission aided the research with data and suggestions. I am especially grateful to George Mistriotis, a programmer at the National Energy Board, for his assistance.

In the three years since the thesis was accepted, many individuals have listened, argued, cajoled, and generally helped in the transformation into what is, I hope, a readable book. To attempt to name all those who gave useful suggestions at seminars or in reading various drafts would take up a good deal of space. Two should be singled out. Rod Dobell made many suggestions, some of which I have incorporated. Tom Russell at the University of Essex, in reading a near final draft, uncovered a vital flaw. Responsibility for any remaining errors must remain with me.

Earlier drafts of the model and results section were given at the Canadian Economics Association meetings, St John's, Newfoundland, and at the European Econometric Society meetings in Barcelona, Spain. Two sections of the book have appeared in altered form. 'National Policy and Natural Gas: The Costs of a Border,' which appeared in the *Canadian Journal of Economics* (August 1972), gave a short description of the model and its major findings. In the *American Economic Review* (September 1972), an extension of the dual chapter appeared ('The Preventive Tariff and the Dual in Linear Programming').

Finally, thanks to Mary Wood at the University of Essex, who typed at least two versions of the book.

This book has been published with the help of a grant from the Social Science Research Council of Canada, using funds provided by the Canada Council.

Brightlingsea, Essex November 1971

Natural gas and national policy

Introduction

The development of the oil and gas industry in North America has been relatively recent. Major discoveries were first made in the 1920s and a great deal of industry growth has occurred since the second world war. Although relatively young, the industry faces a large number of restrictions and regulations imposed by varying levels of government. Rightly or wrongly, governments have singled out the industry as having special significance for public welfare. Many of the restrictions are aimed at preventing the pattern of development which would result from unregulated laissez-faire.

Among the most important of the restrictions are those which prevent free trade between Canada and the USA. Both countries in the past were reluctant to allow unregulated exports and imports. To Canadian authorities energy represented both the road to industrialization and the means to cement relations between western and eastern Canada. While national security was mentioned as a possible excuse for American reluctance to trade, restrictions on imports and exports were an important ingredient in maintaining the profitable position of American oil and gas producers.[1]

1970 has seen increased debate on the possible benefits of removing

1 / That the American producers have benefitted is fairly obvious. M. A. Adelman (1965), Stigler (1971).

National security does *not* mean that imports should be prevented. If a country is concerned that foreign supplies would be cut off in a war, surely the most sensible policy is to save *all* domestic production for emergencies and to use only imports for normal day to day needs.

barriers which restrict the flow of energy resources between Canada and United States. Much of the discussion has, however, centred around the sole issue of increased Canadian exports to the USA.

This book attempts to estimate the additional costs paid by final consumers of natural gas because of restrictions on trade between Canada and the USA. In addition, the benefits of trade restrictions to producers are measured. As these two estimates are but ingredients in the discussion of increased trade today, in this introduction I will state clearly what this study can and cannot tell us and its limitations.

For a single typical year (1966) two models of gas flows in North America are developed using linear programming tools. The estimate of supply is taken to be the actual net production in nineteen aggregate supply areas in 1966. Demand is assumed to equal the actual 1966 net consumption in nineteen aggregate consuming areas. The first model (the free trade model) allocates supplies to demands so as to minimize the costs to the final consumer, where no restrictions are placed on shipments across the border. The free trade model yields as its solution a hypothetical flow pattern which would have existed in 1966. A second model (the constrained model) incorporates a set of restrictions which prevent American penetration of Canadian markets. In the solution to this constrained model, the actual natural gas flow pattern of 1966 is simulated. Comparisons are then drawn between the free trade and constrained models in order to observe the added costs of trade restrictions borne by final consumers and the distribution of these costs among consuming areas.

Each of the two models described in the previous paragraph has an associated dual model which yields information on the prices (shadow values) at supply points and at markets. These two dual models are analysed so as to determine the impact of trade restrictions on prices. The investigation of changes in field prices indicates the benefits of trade restrictions to producers. The dual models are developed further to indicate how they give information on the comparative disadvantage of domestic suppliers in domestic markets. As a result, necessary preventive tariffs (tariffs sufficient to prevent entry) are shown to be combinations of dual prices.

The linear programming method uses a number of simplifying and restrictive assumptions. While these are elaborated in chapter three, let me indicate their nature here. First, the free trade model which is used as a comparison with the actual flows is hypothetical, it would not exist. The free trade model determines a flow pattern for 1966 were producers able to reallocate their production to market demand, given the costs of production and transportation but assuming no existing pipeline network. Of course, producers in 1966 could not have reallocated production in this way since they would be forced to use existing pipelines. A complete analysis of how free trade would affect the distribution of natural gas in 1966 would involve lifting restrictions in some previous year (1950–60) to allow time for adjustments. Then for each year up to 1966, a distribution pattern would be determined, as would a set of prices at markets and at supply points. These prices would themselves be introduced into demand and supply functions so as to determine the levels of production and demand in subsequent years. Comparing the pattern of 1966 trade flows resulting from the complete application of this dynamic model with the actual flows for 1966, would yield true estimates of the costs of trade restrictions.

A complete model is not given in this book. No attempt is made to determine how free trade would affect prices and thus demand, exploration, and supply. The simple model used in this book will yield lower estimates of the inefficiency costs of trade restrictions than a complete dynamic model.

This study offers estimates of the costs to consumers and the benefits to producers of restricted trade. It does not and cannot attempt to measure the social benefits which self-sufficiency has been said to yield. The values of strengthening east-west ties in Canada and the avoidance of foreign sources of energy for the USA are not measured. Instead this study places a value on the costs of these policies. Politicians, given these cost estimates, must answer two questions. First, are the social benefits conceivably of the same magnitude as the costs? Secondly, given that these are worthwhile goals, are there simpler, more direct, and less costly means of achieving them?

Finally, as will be shown, certain producers do gain from trade restrictions. Rather than attempting to fall back on social clichés, perhaps these profits are good alternative explanations for the presence of restrictions.

1 The growth of regulation

1 INTRODUCTION

The oil and gas industry of North America is unique in its position of regulation and protection. No other ostensibly competitive industry is faced with such a large number of restrictions implemented by different levels of government.[1]

The production of oil, and to a lesser extent gas, is controlled by state and provincial conservation boards. These conservation boards, established to reduce the waste characterizing the depression years, use prorationing policies to control the volume of oil produced (and of gas, since 20 to 30 per cent of gas is found dissolved in oil) (Adelman, 1962), and thus the price. In addition, the Conservation Board of the Province of Alberta follows a practice of requiring producers to maintain an inventory of fuel for the future needs of consumers in that province (exportable surplus policy).

Federal authorities in both countries to a great degree control industry developments. In Canada the National Energy Board is the watchdog, approving the construction of all pipelines. Canadian oil policy requires that all oil consumed west of the Ottawa valley be produced in Canada. Canadian gas policy requires all imports and exports of that energy to be

1 / Other industries which are regulated to an even greater degree are government chartered monopolies – telephone service, electricity distribution, etc.

approved by the National Energy Board. Gas is not allowed to be exported until the Board is convinced that Canada's future needs have been adequately provided for (exportable surplus policy). In order to prevent imports of American gas into eastern Canadian markets, the Canadian government subsidized the construction of an intercontinental gas line, wholly within Canada (Trans-Canada Pipeline). American authorities similarly restrict imports and exports. Oil imports from all countries but Canada face quotas, while Canadian supply is limited by moral suasion. The Federal Power Commission, given jurisdiction over natural gas pipelines in 1938, must approve all exports and imports of gas. The FPC also regulates the prices which American gas producers can charge in inter-state commerce.

At one time or another, officials of gas producing and consuming regions in both countries have spoken against trade. In addition, the federal authorities in both countries have rejected both import and export licences.[2]

One can well imagine that the haphazard growth of regulation, formed by agencies on several levels of government and on both sides of the border, has not resulted in a consistent pattern aimed at benefitting the interests of any one group. However, it will be shown that, in fact, most policies are complementary. The policy of the FPC in delaying or rejecting licences to import Canadian gas buttresses the reluctance of Canadian authorities to allow exports.

If the policies complement each other, what group in society is better off because of trade restrictions? Regulations and restrictions were not introduced with the explicit wording in houses of legislature to make this person or that group better off. Instead, the imposition of regulation has been couched in terms of 'the public interest,' or 'national defence.' When the Federal Power Commission refused to permit a licence to import Canadian gas in 1953, the authorities suggested that American consumers could not be dependent on producers in another nation whose authorities might request them to turn off the tap. American oil quotas are praised

---

2 / In 1951, the FPC refused to allow the export of gas to Canada. In the middle 1950s, it rejected several applications to import Canadian gas to the USA. In 1953, the Canadian government refused to allow imports of gas. The policy of the Canadian government is to hold hearings on all export applications for gas, several of which were refused in the 1950s, and several which may be rejected this year.

by oil producers as contributing to the ability of the USA to withstand an emergency which shuts off foreign supplies.[3] Canadian opposition to trade is also based on security grounds, and on a long-standing national policy to promote east-west trade.

It is simple for specific groups to confuse their self interest with the public interest. Indeed one would not expect interest groups to be overly concerned with the welfare of the general public. Regulations or trade restrictions which are introduced by pressures from producers are least likely to be in the 'public interest.'

One can naïvely accept the view that the regulation of industries is introduced either by powerful consumer opinion or by altruistic producers seeking the best interests of the nation.[4] Alternatively, one can view regulation as an attempt by industries to use the coercive power of the state to improve their lot. '... as a rule, regulation is acquired by the industry and is designed and operated primarily for its benefit' (Stigler, 1971, 3). This latter view is, in my opinion, the easier and more rational view to accept. The view of regulation as being primarily in the public interest suggests that society is able as a group to establish its goals. The body politic sees that firms acting alone to maximize profits would not meet certain social goals. Society then coerces firms to behave in a way inconsistent with profit maximization. This view suggests that society can set collective goals and that decisions taken in legislatures reflect these goals. Recent literature calls these two assumptions into question. The process of collectively arriving at a single unambiguous goal appears theoretically impossible to make in a democracy (Arrow, 1963). Analyses of our actual democratic process suggest legislation is introduced as trade offs among interest groups for favours rather than as the will of the people reflected by their elected representatives (Arrow (1969), Buchanan and Tullock (1962), Tullock (1970)).

The actual developments of regulation of energy in general and gas in particular does suggest that certain regulations were introduced to aid

---

3 / 'The "protection of the public" theory of regulation must say that the choice of import quotas is dictated by the "concern of the federal government for an adequate domestic supply of petroleum in the event of war" – a remark calculated to elicit uproarious laughter at the Petroleum Club' (Stigler, 1971, 4.)

4 / Kilbourn in his book on the trans-Canada pipeline (1970) shows an immense naïvety in accepting at face value that an American owner of gas reserves pressured for the trans-Canada pipeline for the good of Canada. See Waverman (1971).

industry members while other restrictions do not appear, at least at first glance, to be in the self-interests of producers. As the later sections of this book suggest, the total effect of a number of restrictions, some of which do not themselves benefit producers, may however be to increase the earnings of producers.

The hypothesis of this book is that, in general, producers in Canada and the United States are better off, while ultimate final consumers are worse off, because of the restrictions on trade in natural gas between the two countries. The estimated present value of the increased costs to consumers because of the actual inefficient transportation pattern in North America is some two hundred million dollars. Most of this increase in costs is borne by the consumers in eastern Canada. Producers in western Canada earn higher profits under trade restrictions as opposed to free trade.

2 THE ORIGINAL NEED FOR REGULATION

Regulation is necessary in the oil and gas industry since the production of these minerals in an unregulated competitive world leads to waste. This waste in competition occurs because of the presence of technological external diseconomies in production. An increase in the number of wells or in the rate of production for a single producer can increase the costs for other producers.

Gas and oil are found in pools, large areas of mineral-bearing rock which will usually be under the land of a number of different owners. The gas or oil in the pool is a homogeneous commodity, any two units from any two sections of the pool being identical. It is difficult to determine the share of the entire pool which lies under each individual's land. An individual can therefore drill many wells on his single plot of land and capture a very large share of the gas or oil in the pool.[5] It is in the best interests of every owner in the pool to maximize his present production, for what is not produced today may be produced by others tomorrow. Production is maximized by maximizing the number of wells and the production rate

---

5 / The courts have held that flowing oil and gas have the properties of migratory birds. As a result, ownership is determined by the law of capture. If you catch it (bird or oil) you own it. Possession is 100 per cent of the law.

for each well. An individual's decision to so maximize his production reduces the pressure in the pool for other owners. At some point this reduction in pressure decreases the flow from other wells. Oil and gas production are characterized by technological externalities – a divergence between private marginal costs (the costs an individual producer considers) and social marginal costs (the costs to society for one more unit of output). In Figure 1.1, *Sp* represents the supply curve for a pool under unregulated perfect competition. *Sp* is the sum of all the individual marginal cost curves. A competitive industry would be in equilibrium where this supply curve (*Sp* cut the demand curve *DD*. Competitive output is *Qp*, price is *Pp*. The true social costs of an output at level *Qp* is not *Pp*, but *Po*. *Po* – *Pp* represents the costs imposed on other producers from the last unit produced in the unregulated market. *Po* – *Pp* is the cost not taken into account in the individual firm's decision, the cost which is borne by other producers. Equilibrium should be at an output of *Qs*, with price in the market of *Ps*, at the intersection of demand and social supply (*Ss*).

Unrestricted competition in the production of oil and gas leads to over-

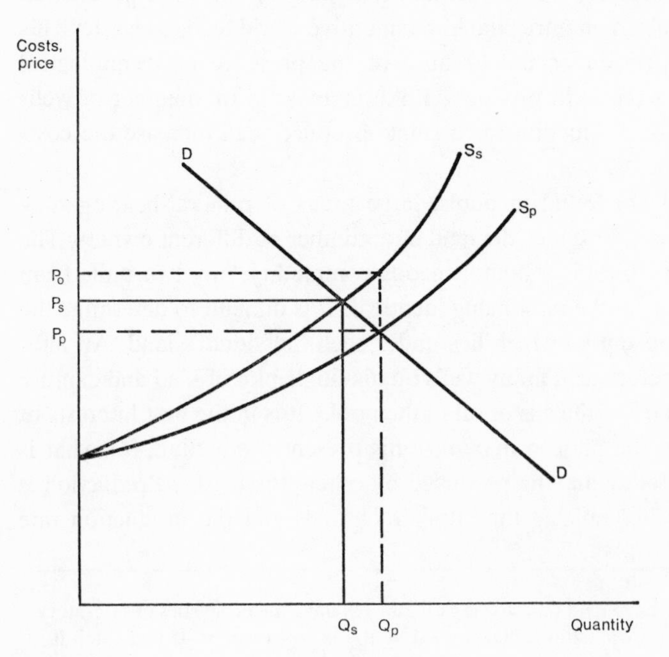

FIGURE 1.1

production ($Qp > Qs$) at too low a price ($Pp < Ps$). Regulation is one possible method of preventing this wasteful competition.[6]

Conservation policies were first established to reduce the overproduction characterizing unregulated *laissez-faire*. By establishing well-spacing rules (the minimum acreage per well) and the maximum efficient capacity of a well (MEC – the production rate beyond which other producers are hurt), regulation ensured that external diseconomies would not occur. This form of regulation improves the lot of both producers and consumers – all benefit.

Once the regulatory body collected data on each well – its location, its maximum efficient capacity and actual production, the possibility existed that wells could be forced to limit production even further. Conservation to avoid technological diseconomies becomes prorationing aimed at maintaining or raising the price.[7]

3 CANADA – PROVINCIAL REGULATION

Until 1949, all oil wells in Alberta were able to sell their maximum efficient capacity. Because of large discoveries brought to production in 1949 and 1950, '... production potential increasingly outstripped the quantity that could be marketed' (Hansen, 1958). In Figure 1.2, the industry is represented at equilibrium at a quantity sold Q, at a price P. Supply increases from SS *to* S^1S^1. The new equilibrium is at a higher output Q^1, and a lower price P^1. Supply does not 'outstrip' demand unless producers refuse to lower the price from level P. With an increased output at a price of P, excess supply of *cd* exists.

The oil companies suggested to the Alberta government that it had the authority to establish quotas for each field, to prevent 'excess' production of type *cd* (so that price would not fall). Conservation boards

6 / Several other methods can be used to force individual firms to recognize the true social costs of their actions. Unitization, used in Colorado and Saskatchewan requires all pool owners to make a common drilling and production plan, recognizing that a single owner of the pool would not increase the number of wells in the northeast corner merely to reduce production from established wells in the southwest corner. Taxation is another method of reaching the socially best outcome.

7 / Lovejoy and Homan, 1967. The authors hedge their conclusions but do suggest that market prorationing maintains the price. Also see Adelman (1965).

today in Alberta and British Columbia base their quotas for wells on the estimates of demand, *given the price.* Market prorationing is an example of Stigler's maxim of regulation for the benefit of producers. Without a government programme which coerces each producer to cut back his production, on threat of penalty, competition would lower the price benefitting consumers.[8]

While only oil is directly prorationed, this policy does affect the supply of gas, since nearly one-third of gas is found dissolved in oil.

The 1938 bill establishing the Alberta Conservation Board gave the Board the authority to approve all exports of hydrocarbons beyond the boundaries of the province. In 1949, the first request to export gas out of Alberta was rejected by the provincial Oil and Gas Conservation Board. The Board stated that there were insufficient gas reserves in Alberta to meet thirty years' future requirements of consumers. This policy of requiring producers to maintain an inventory of proved reserves

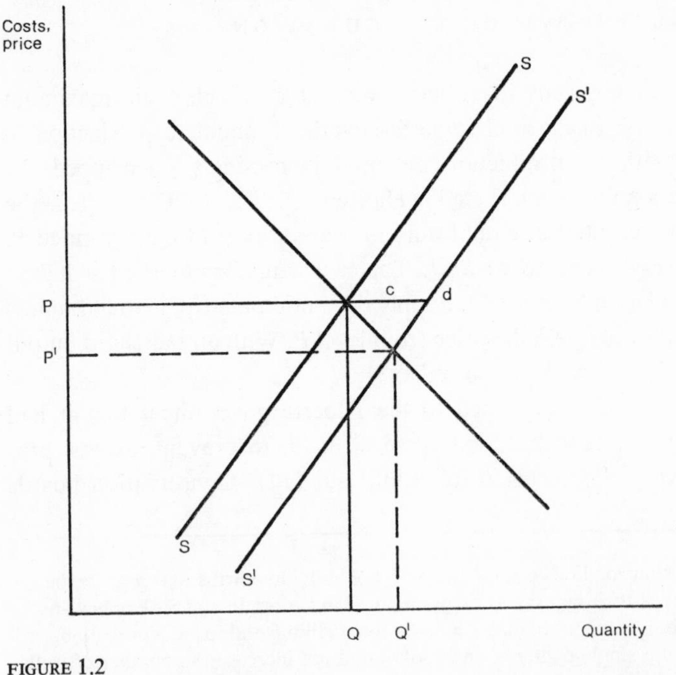

FIGURE 1.2

8 / Market prorationing was introduced to eliminate 'excess competition' and for 'national security' reasons. Adelman (1964) estimates that the US government could end prorationing, buy out all the small marginal wells (stripper wells) in the USA for $1.12 billion, and still save consumers $2.5 billion annually.

on hand for domestic consumers (exportable surplus policy) is a major ingredient for Canadian federal authorities also.

The exportable surplus policy makes implicit assumptions about the trend in future prices. Future consumption is estimated for the next thirty years. Present reserves are known. If reserves are greater than this demand estimate, the 'surplus' can be exported.[9]

As in the policy of market prorationing, the concept that demand is a schedule, whose exact value depends on the price is not explicitly stated. The reserves necessary to supply domestic consumers for the next n years is obviously a function of price for two reasons – demand being determined by the price charged and the amount of gas supplied by producers being price determined.[10]

Producers are thus constrained to maintain an inventory of gas in the ground which will be sufficient to meet domestic demand for n years and keep price constant. Since producers would seem to prefer selling this gas to the highest bidder, whether in the province, in the country, or in the USA, an exportable surplus policy by itself does not improve the welfare of producers.

4 CANADA – FEDERAL REGULATION

Exportable surplus

In 1951, the Alberta Conservation Board decided that an exportable surplus of gas existed. An agency of the federal government, the Depart-

9 / Important to the policy is the method of determining reserves. The National Energy Board has through time liberalized its notions of reserves so as to increase the probability of export. Percentages of 'probable' reserves (reserves not established but discovered), have been added. In 1970, the Board included for the first time in its estimate of reserves 50 per cent of reserves then beyond economic reach (too far away to be commercially exploited in 1970). See Hamilton (1971).

10 / Future demand for Canada has been estimated by the National Energy Board on three bases, each of which implicitly assumes the price of gas to remain constant or change in the future as it has in the past. The three methods used (at least as reported through 1967) do not take price into account. First, trends in energy consumption are extrapolated to the future, assuming the same growth rate as in the past. Second, energy consumption is related to GNP for the past. Future energy requirements are then estimated from exogenously given GNP forecasts. Finally, the end uses of each fuel are determined and the 'needs' of these aggregated end users estimated.

ment of Trade and Commerce, had the power, under the 1907 Exploration Act, to authorize the export of gas from Canada. Another federal agency, the Board of Transport Commissioners, had jurisdiction over the route of any interprovincial or international pipeline. Faced with an application to export Alberta gas to the USA in 1953, the federal government announced its desire to maintain western Canadian gas for eastern Canadian consumers. '... the policy of the government of Canada is to refuse permits for moving natural gas by pipeline across an international boundary until such time as we are convinced that there can be no economic use, present or future, for that natural gas in Canada.'[11]

An exportable surplus policy for an energy-producing area such as the province of Alberta benefits consumers in that province because of the tendency to restrain price.[12] An exportable surplus policy for an area as vast as Canada need not benefit consumers in Canada generally. In the province of Alberta, provincial producers have a natural advantage over foreign producers since the costs of shipping gas into Alberta from other sources is very high. Policies, then, which prevent provincial producers from taking advantage of their location in pricing to provincial users, benefit consumers. An exportable surplus policy for a nation as a whole benefits consumers of that nation if it prevents domestic producers from taking advantage of their location. A policy aimed at forcing western Canadian producers to maintain a twenty-five to thirty years' inventory of gas on hand benefits consumers in eastern Canada, if these markets would have been supplied by Canadian producers without this regulation. In that case, regulation tends to lower the price, and disregarding the effects on exploration and future supply, improves the welfare of consumers in eastern Canada. When an exportable surplus policy is combined with the explicit restriction of imports, however, consumers may be worse off.

In fact, imports had to be forbidden from entering eastern Canadian markets. In 1953, Consumers Gas Co of Toronto, Ontario requested approval of an 80-mile link to Buffalo, New York. An American company (Tennessee Pipeline Company) would supply Toronto with gas. The

---

11 / C.D. Howe, before the House of Commons, March 13, 1953, quoted in Kilbourn (1970), 35, 36. This statement was made while a Canadian gas transmission company was before the Federal Power Commission for an import licence.

12 / For example, in Alberta cheap gas is considered a political necessity, '... the right to a free fuel supply in many rural areas was considered as natural.' *Ibid.*, 18.

federal authorities did not have explicit authority to restrict imports. The application for imports was denied under the authority of the Navigable Waters Protection Act. That Act, which forbade impediment to shipping, was used to prevent construction across the Niagara chasm.[13]

The all-Canadian pipeline

A policy which rejected exports until 'all Canadian demand present or future had been provided for' and the restriction of imports may have been necessary policies to promote a west-east movement of gas within Canada. These policies were not sufficient, however, and the government itself was forced to assist in building a transcontinental pipeline. The Liberal government of 1955 was not content with a pipeline which just joined western gas producers and eastern consumers. The line had to be built wholly within Canada. This policy of a gas pipeline constructed entirely within Canada was a complete shift of the government's policy enunciated in the construction of the inter-provincial oil pipeline six years earlier. In 1950, C.D. Howe, the man most responsible for the all-Canadian route of the gas pipeline, was opposed to a route only within Canada for the oil pipeline from western to eastern Canada. 'But Howe, the practical engineer and builder of business, was not one to be moved by histrionics. He stood firm by his decision to allow the more economical US route' (Gray, 1970).

By 1955, Howe and the Liberal government were committed to build a natural gas pipeline north of Lake Superior. The building of the trans-Canada pipeline in 1956 involved two decisions – one to connect eastern and western Canada, the second to do so via a route within Canada. The first decision is fairly easy to explain as it has been made many times before and after 1956. This Canadian policy is based on the ancient theme of Canadian development, the construction of east-west trade routes in the face of dominant north-south flows.

Canadian national policy, the conscious attempt of creating a viable political and economic unit, arose in the early 1850s. The repeal of the British corn laws in 1846 and the difficulties in establishing reciprocity with the USA convinced leaders of the British North American provinces

---

13 / A boat took tourists close to Niagara Falls, a technicality which provided the excuse to use the Act.

that development depended on the strengthening of ties between the scattered provinces. By the time that national policy was explicitly formulated in 1879, a philosophy was completed, a viable political unit necessitated economic integration through the mechanism of an east-west trade flow.[14] The policy of promoting east-west trade flows within Canada for gas and oil is consistent with a central theme of Canadian development. The second decision, the desire to build the gas pipeline entirely north of the border, parallelled a similar decision taken eighty-eight years earlier.

In 1867, large gaps remained in the Canadian transcontinental rail system. Upon entering confederation, both the Maritimes and the West were promised rail links. The British vetoed the cheapest route to the Maritimes, which cut through northern Maine, on the grounds that a military transport system could not cut through foreign lands. To the west the most efficient proposal was to use existing American lines to a point south of Winnipeg and to construct a new line from Winnipeg to the Pacific. The proposal was rejected. 'The railway, after all, was not intended to be merely a method of transportation of goods from one place to another, it was also an instrument of national unification. How could it serve this purpose if the vital link between east and west remained under the control of a foreign power?' (Easterbrook and Aitken, 1958). It appears as if these thoughts guided the Canadian federal government in the nineteen-fifties and -sixties. A pipeline is not, however, 'an instrument of national unification.' It has but one purpose – to ship a homogeneous commodity identical to that which could be imported. Moreover, at the time of its construction, the Trans Canada Pipeline Company was controlled by American interests, as was (and is) the production of oil and gas and the refining of oil.[15] It is therefore questionable whether the trans-Canada pipeline was an important ingredient in Canadian national policy.

The Trans Canada Pipeline Company experienced great difficulties in financing construction.[16] In early 1956, the federal government agreed to

---

14 / For an excellent (and short) description of the development of national policy, see Fowke (1952).

15 / Canadian trade minister Pepin is quoted as saying in the House of Commons that 82.6 per cent of the oil and gas wells industry was American owned as was 99.9 per cent of oil refining (Laxer, 1970, 17).

16 / The problems of financing are quite similar to those described for the Union Pacific Railway in the USA (Fogel, 1962). Basically, in both cases, investors did not believe that the line would be completed and be profitable. In the case of the trans-Canada pipeline company, investors must have doubted both the ability to construct

build by itself the most difficult section of the line from the Manitoba-Ontario border to north of Lake Superior at Kapuskasing, Ontario, a distance of 676 miles.[17] The Company had to construct 1618 miles of pipe, most of it through the Prairies; it was still unable to raise sufficient financing, and, in April, 1956, asked the government for further assistance. Amid tumult in the House of Commons, the government passed a bill on June 6, 1956, which in addition to the section to be built by the government, lent the company up to $80,000,000 for five months at 5 per cent.[18]

A very rough estimate of the increased construction costs resulting from the northern route can be arrived at in the following way. The government built 676 miles of pipeline (including compressor stations) at a cost of $129,856,000.[19] The company built 1618 miles at a cost of $245,000,000[20] (including compressor stations). The cost per mile of the section through northern Ontario was $192,095. The cost per mile for the remainder of the route was $151,420.[21] These figures are not directly comparable since the prairie section was 34-inch pipe while the northern Ontario link was 30-inch pipe. Also, the compression per mile may have been different. Based on these figures, the 676 miles built north of the Great Lakes cost $27,500,000 more than an equidistant line through prairie soil.

The second debate on an all-Canadian route

By 1963, increased demand in eastern Canadian markets made a second pipeline to the east possible. Great Lakes Transmission Company, a joint

---

a line through muskeg and the determination of the Canadian government to maintain eastern Canadian energy markets for western Canadian suppliers. The appropriate decision would have been for the government to construct and operate the line, for the two uncertainties bothering private investors were sure things for the central government.

17 / The Ontario portion of the line was leased by the Trans Canada Pipeline Company for five years until purchased at original cost in May 1963.

18 / The bill was passed by invoking closure on all four stages of the bill, the first time such a technique was used in the Canadian parliament. This bill is thought to have led to the defeat of the Liberal party in the election of 1957.

19 / *Trans Canada Annual Report* (1963).

20 / Trans Canada Pipeline Company news release (Feb. 2), 1959.

21 / The laying of feeder lines shows the difficulty of construction in rock. Five contracts for a total of 367 miles of 30-inch pipe totalled $20,827,789 or $10.10 per foot compared to $3.50 per foot in the prairies (*Oil and Gas Journal,* no. 49, Dec. 9, 1957, 65). The difference is over $35,000 per mile.

subsidiary of the Trans Canada Pipeline Company (the all-Canadian route) and American Natural Gas, applied for permission to construct a 1000-mile pipeline between Emerson, Manitoba and Sarnia, Ontario, wholly within the USA. The line would deliver 507 MCFD for sale in the American market, on its way to supplying eastern Canada.

In early 1966, after three years of deliberation, the National Energy Board approved the plan. On August 25, 1966, Prime Minister Pearson, speaking for the cabinet, announced that the government had rejected the National Energy Board's approval, stating: 'The government does not believe it is to be in Canada's best interest that the future development of facilities for bringing western gas to its eastern Canadian market should be located outside Canadian jurisdiction and subject to detailed regulation under the laws of the United States which are naturally designed to protect the interests of the United States citizens.'[22]

Later that year, the Canadian cabinet reversed its August position and approved the project, as amended. The amendment stated that over 50 per cent (60 per cent by 1976) of western gas reaching eastern Canadian markets must move through the all-Canadian route. In order to abide by this amendment, Trans Canada Pipeline Company agreed to loop the section of the existing line in northern Ontario.

Tariff policy

The protective tariff has been an instrument of national policy since 1878. The federal government imposed a tariff of six cents per MCF in 1924, later lowering it to three cents per MCF. The Kennedy round of GATT eliminated the Canadian tariff on natural gas as of January 1, 1968.[23]

There is evidence that this high tariff was insufficient to prevent American gas from entering Canada. In 1955, Consumers Gas of Ontario again requested permission to import gas from the USA for sale in the Ontario market (after paying the tariff). The Department of Trade and Commerce suggested that Consumers Gas wait until the construction of the trans-Canada pipeline when gas would be available at comparable cost. This policy was not formally enforced because it would have been

22 / *Montreal Gazette* (Aug. 27), 1966.

23 / Chapter 6, which examines tariff policy, shows that the three-cent tariff was totally ineffective in preventing American entry. Its elimination was not a great loss to Canadian authorities.

difficult for Canada to impose import restrictions when she was trying to persuade the USA that they were infeasible.[24]

An alternative proposal to the Great Lakes transmission line of 1963 was presented by Northern Natural Gas. Northern proposed to save fifty-five million dollars on capital costs by an interchange of American and Canadian gas, American gas going to Ontario users and Albertan gas going to the American midwest.

One of the main arguments against this alternative when it was rejected in 1967 was that it would not move the gas in bond and therefore would be liable for Canadian import duties of $5,475,000 per year. This proposal was not re-examined after the tariff was eliminated.

After the tariff was removed, direct prohibition continued. In 1968, Union Gas (Ontario) which was receiving 65 MCFD from an American supplier, Panhandle Eastern, asked for permission to receive an additional 100 MCFD by 1974. In February 1969, the National Energy Board rejected this proposition.

Canadian policy – summary

Canadian authorities have used a wide range of policies in order to promote a west-east flow of gas within Canada. A tariff was in effect for forty-four years. Perhaps, when it was recognized that the tariff was not a real barrier, importers willing to pay the tariff were refused licences. Besides curbing imports, the use of an 'exportable surplus' policy has restricted exports. It is obvious then that free trade on Canada's part has not been an important ideal in the natural gas industry.

5 UNITED STATES – STATES REGULATION

The rise of regulation in the USA parallelled the growth north of the border. Producing states were the first to regulate oil and gas shipments, with the federal authorities intervening in the 1930s.

The great increase of productive capacity in the depression years of the

24 / '... Moreover while natural gas is imported into Ontario from the United States, in large part such imports are on a temporary basis. Thus, in effect, the increasing demand for natural gas in Eastern Canada must be met from Alberta sources' (Royal Commission on Energy, *First Report*, Oct. 1958, 1–11).

1930s convinced many states (Texas and Oklahoma principally) to introduce some scheme of prorationing production to demand. Today, 75 per cent of the oil produced in the USA flows from states which use market demand prorationing (Lovejoy and Homan, 1967). No state, however, has used the concept of the 'exportable surplus' to limit the export of oil or gas beyond state boundaries.[25]

6 UNITED STATES – FEDERAL JURISDICTION

Prices

In 1938, after several courts had disallowed the right of individual states to set the price at which natural gas moved in inter-state commerce, the Federal Power Commission was given the authority to regulate inter-state pipelines. The Natural Gas Act of 1938 authorized the FPC to regulate pipelines as natural monopolies and to set a fair return on their rate base. Since a large portion of the price of natural gas delivered at the market is the price paid by pipelines to producers, the FPC soon became involved in a controversy over whether it had the authority to regulate the price at which gas was sold to pipelines. The Commission disavowed any interest in regulating the prices paid by pipelines (field prices). In 1947, the Supreme Court held that the FPC did have jurisdiction over gas producers who sold in inter-state commerce. First, by attempting to frame a bill before the House and second by establishing an internal order, the Commission attempted to divorce itself from the regulation of independent producers who owned no pipelines.[26]

The gas-producing industry marshalled its forces to press for amendments to the Natural Gas Act which would exempt independent producers from regulation. Between 1947 and 1958, seven bills were introduced into congress to exempt gas producers from federal regulation.[27] Two

25 / Since for an oil-producing area, an exportable surplus benefits consumers at the expense of producers, it is possible to suggest that producers have more political power in American than Canadian petroleum-producing regions.

26 / Statement of the FPC before the House Committee on Interstate and Foreign Commerce (June 23, 1947). Order no. 139 (Aug. 7), 1947.

See Nash (1968) for a description of the attempt by the FPC and producers to prevent price regulation.

27 / Pierce Bill (July 8, 1947); Kerr Bill (April 4, 1949); Fulbright Bill (June 6,

were passed (Kerr Bill, 1950; Fulbright-Harris Bill, 1956) but were vetoed by the president. In 1954, the Supreme Court reaffirmed its 1947 decision, in effect ordering the FPC to regulate the prices paid to all interstate producers.[28] The FPC accepted its mandate, and set prices for producers by establishing ceilings on prices for large producing areas (area rate setting).

The continued attempts by the industry to forestall regulation of producers does not suggest that producers felt they would gain from this regulation. Various studies have shown that producers have not benefitted (Gerwig, 1962). MacAvoy states that consumers also have been hurt – the benefits of regulation may be less than its costs. MacAvoy also suggests that the establishment of ceiling prices may have induced producers to restrict exploration and discovery (MacAvoy, 1971).

USA federal policy – import and export

The attitude of the American authorities to the importation of oil is well documented. Quotas for most of the world and implicit restrictions on Canadian imports have raised the costs to consumers and protected producers. In the natural gas sector, the Federal Power Commission has the authority to approve all exports and imports. While American trade restrictions on gas flows do not appear to be significant now, there have been occasions in the not-so-distant past when both exports and import permits were rejected. As gas first moved north from fields in Texas to the northeastern states, local Canadian natural gas distributors began to consider importing this gas. In the late 1940s, Consumers Gas of Toronto began discussions to import gas from Buffalo, eighty miles away. At the same time, consumers in northern New York state objected to the possibility of such exports to Canada (Kilbourn, 1970, 15).

In the early 1950s the FPC refused *both* export and import licences. In 1951, an export licence application was rejected, on the basis of the gas being needed in the USA (an exportable surplus theory). However, in 1953, the FPC did reverse its earlier decision and allowed Tennessee Gas Transmission to export into Ontario. Whether or not the USA would have

---

1955); Harris Bill (July 28, 1955); Fulbright-Harris Bill (Jan. 31, 1956); Harris Bill (March 14, 1958); Harris Bill (March 7, 1960).

28 / *Phillips Petroleum Company* v. *Wisconsin*, 347 US 672 (1954).

allowed exports as producers wished is an academic question, for the Canadian policy-makers quickly showed that such exports into Canada were not acceptable. The first approval of gas imports to the USA involved a small Korean war emergency service to a copper mine in Montana.

In October 1950, a Canadian company, Westcoast Transmission, applied for permits from Canadian and American authorities for service from northern Alberta to the northwestern United States. In June 1954 the FPC denied Westcoast permission to build a 400-mile pipeline. Instead, Pacific Northwest Pipeline Corporation was permitted to build a 1400-mile line to these states from the San Juan Basin in New Mexico and Colorado, a natural gas reserve which even in 1954 was inadequate to serve California and Washington.

This decision to restrict imports of gas into the USA has been interpreted in a number of ways. Several authors (including a Canadian Royal Commission), have concluded that this decision was but one of the many judgments in the energy field forcing the US to be self sufficient. National security required that vital materials be in sufficient supply to ensure adequate use in an emergency. Since natural gas is difficult to store, the FPC attempted to rely as little as possible on foreign suppliers over which it had no jurisdiction.[29]

While there was probably some note of protectionism in this policy, the acts and views of Canadian authorities are probably more responsible. Not only had a royal commission been instituted in Canada to conduct an inquiry into the role of imports and exports, but elected representatives were making public statements as to the need for export controls (see C.D. Howe's remarks quoted earlier, p. 14, n. 11). This was not an atmosphere conducive to trade. Furthermore, Westcoast Transmission had neither received explicit governmental approval nor confirmed gas supplies before applying to the FPC. It is quite remarkable that the FPC considered Westcoast's application for as long as it did, considering political attitudes in Canada and the lack of guarantees of gas (Aitken, 1959).

---

29 / '... the FPC was dubious about having a part of the United States supplied from a region in which it had no jurisdiction over production and transportation' (Hansen, 1958).

The Royal Commission on Energy places sole responsibility for the decision with the FPC. They stated that the USA would never depend on imports from a foreign country 'without some intergovernmental agreement assuring the continued adequacy of supply' (*First Report*, 1958, 6–23).

Westcoast Transmission began negotiations to bypass the Pacific Northwest market and build a pipeline to California. The negotiations had reached the point where a price had been established between the companies involved when, in 1955, FPC permitted imports to the northwestern USA by Westcoast Transmission as a supplementary supplier. In October 1957, seven years after the initial plan and five years after Westcoast had received approval from the Alberta Oil and Gas Conservation Board, the line was completed.

In April 1960, the Trans Canada Pipeline Company received permission to import into the American midwest, only two years after the FPC had initially rejected the application.

Late in 1957 a plan was announced by a consortium of Canadian and American companies to build a 1300-mile 36-inch pipeline from Alberta to California. After four years of negotiations with the Canadian and American authorities, the Alberta and Southern Gas Company began operation.

In the late 1960s, the FPC approved all applications to import gas into the USA from Canada. In 1970, worried over a growing shortage of gas in the USA, the FPC called for increased exports from Canada.[30]

7 SUMMARY

Until recently, the policies of the two countries have complemented each other. Canadian policy, based on a theory of exporting only 'surplus' gas is aimed at eliminating imports and restricting exports. In the 1950s, American officials, either for national security or other reasons, rejected applications to both export gas from and import gas to the USA. In addition, the costs involved in the long time lags of the regulatory process – first gaining approval from Canadian officials then applying to the FPC – undoubtedly raised the price of gas and this tended to limit exports from Canada.[31]

Some policies appear to have been introduced to benefit the general

30 / MacAvoy (1971). MacAvoy finds that the shortage can be explained by the Commission's too low prices in fields.

31 / Gerwig, 1962, estimates that the 'normal' delays for domestic American projects cost some $110,000,000. The delays in international proceedings were four to six times as lengthy as in domestic applications.

consuming public (Alberta's exportable surplus policy) while other policies obviously raise producers' welfare (prorationing of oil for example). The policy of restricting trade between the two countries does not obviously fall in either category of improving consumers' or producers' welfare alone.[32]

At first glance, it would appear that producers would not benefit from a policy which prevented them from selling to the highest bidder, when that bidder is foreign. However, a co-ordinated policy of restricting imports giving producers greater monopoly power in the domestic market may increase their profits. The net result would be an improvement in producers' welfare.

One would expect consumers to be worse off in the presence of the trade restrictions. A quote of C.D. Howe[33] demonstrates that even politicians admitted to the cost of these policies. No magnitude has as yet been placed on these costs. It is the goal of this book to make such an estimate, both of the costs to final consumers and the possible benefits to producers.

---

32 / Free trade maximizes social welfare under a number of assumptions – perfect competition, the absence of externalities, continuous functions.

33 / 'National Policy required that gas surplus to Alberta's needs must not be committed outside the country until Canada's needs had been adequately provided for ... To whatever extent it may be true, let it be remembered that we long ago accepted the fact that there is a price on Canadian nationhood. Had we always sought the cheapest way, there might now be 58 rather than 48 states in the great country to the South, but there would be no Canada.' (Howe, 1955.)

2
The model

1 INTRODUCTION

Taken as given are 1966 production at the supply points (fields), consumption in the markets, production costs and transportation costs. *No* existing pipeline network is assumed. Instead positive flows are permitted from any field to any market. Using this data, a linear programming model is used to choose those flows which minimize the total costs of the system.[1]

This objective function (2.1) (the total costs of producing and shipping a unit of gas from field i to market j) is minimized subject to two sets of constraints. The first set of constraints (2.2) comprises the market constraints. The total shipments into a market from all possible fields must be at least as great as that market's demand. The second set of constraints (2.3) deals with the fields. The total shipments from a single field to all possible markets must be no greater than the total supply available at that field.[2]

---

1 / A complete general reference on the use and solution of linear programming models is Dantzig (1963). A development of the concept in economic terminology is Dorfman, Samuelson, and Solow (1958). Two excellent applications are embodied in Kendrick (1967), and Henderson (1958).

2 / Since the costs associated with a shipment are positive, this equation set will always be satisfied by an equality. If it were satisfied by an inequality, the objective function could be lowered by reducing shipments, yet no demand would go unsatisfied.

$$\text{minimize } Z = \sum_{i=1}^{m} P(x_i) + \sum_{i=1}^{m} \sum_{j=1}^{n} T(x_{ij}) \tag{2.1}$$

$$\text{subject to} \quad \sum_{i=1}^{m} x_{ij} \geq D_j \quad (j = 1, ..., n) \tag{2.2}$$

$$\sum_{j=1}^{n} x_{ij} \leq S_i \quad (i = 1, ..., m). \tag{2.3}$$

x_{ij} = the flow from field i to market j in millions of cubic feet per day (MMCFD); $P(x_i)$ = production costs for quantity x in field i; $T(x_{ij})$ = transportation costs for quantity shipped from field i to market j; D_j = the demand at market j (MMCFD); S_i = the capacity of field i (MMCFD).

2 COST MINIMIZATION AS THE APPROPRIATE OBJECTIVE

The objective function (2.1) minimizes the delivered costs to North American consumers at the market. A number of alternative objectives could be used. For example, the decision rule might be to minimize the return to Canadian producers or to minimize the delivered price in eastern American markets. Each of these three objectives would yield different solutions (different flow patterns and delivered prices).

Some case could be made for any of a number of alternative objectives, depending on whose interests are to be obeyed. The minimization of the costs to consumers was chosen as the objective for a number of reasons. First, the minimization of costs to consumers fits the competitive model. Under perfect competition, costs to consumers (and prices) are minimized. Furthermore, in a competitive model, where consumers costs are minimized, producers profits are maximized.[3]

To single out some segment of producers or consumers and maximize their well-being at the expense of other consumers would necessitate showing why this particular group merits favour. Rather than analyse why every possible alternative objective is inferior to the one actually chosen, I shall instead concentrate on one – that the profits to Canada be maximized.

---

3 / In a competitive industry, producers' profits are maximized at a zero return above normal profits. Chapter 5, which develops the dual to the model (the primal), sections 1–3 shows the analogy between profit maximization and cost minimization directly (72, 73).

A possible objective for Canada would be to maximize the price Canadian producers received in the American market. This suggestion assumes that Canadian producers have some monopoly power in the American market which they are not exercizing. For some reason, Americans are better bargainers, or else, because of political power, are able to get gas at a price lower than the costs of the next best alternative to Canadian supply (Laxer, 1970).

To establish policies which benefit producers in Canada (such as monopoly bargaining with the USA) is not generally a sufficient policy to maximize the interests of Canadians, for it is not at all obvious that the public benefits from increasing the profits of producers of energy in Canada. Unless a tax scheme exists which redistributes these excess profits to the public, producers are better off but not consumers. Since most of the oil and gas industry is American owned,[4] a Canadian policy to maximize producers' profits will just involve a redistribution of welfare from American consumers to American producers.[5]

It is not obvious that a tax policy exists which would redistribute these profits. Nor is it clear that producers in Canada do have unexerted monopoly power in American markets. Therefore, the objective of maximizing producers' profits in Canada is not used here.

Cost minimization for all consumers does not involve trade offs between the welfare of certain groups. Nor does it require a series of assumptions as to the nature of the tax system or the distribution of ownership of wells. Cost minimization is an appropriate objective to study the incidence of inefficiencies generated by public policies.

3 THE COST FUNCTION

The total cost of the activity of transforming gas from a well to a usable product in the home contains many elements. The elements to be included in this model depend on the purpose to which the solution is to be used.

---

4 / See chapter 1, 16.

5 / Maximizing the profits of producers physically within Canada would hurt Canadian consumers also unless two separate price systems were used, the domestic one at minimum costs, the export at monopoly profits. In chapter 6 an export tax policy which has the effect of creating a two-price system is discussed. An export tax system avoids the problem of attempting to tax monopoly profits away from producers.

The purpose behind the exercise is to approximate the behaviour of an actual system were it allowed to relax an operative constraint. Although linear programming is a static single year solution, long-run cost estimates will more likely represent the costs which decision-makers entertained when arriving at the investment policies which resulted in the 1966 flow pattern.

Inter-provincial and inter-state natural gas pipelines are not common carriers. They purchase a commitment of reserves from a producer, normally for a twenty-year period. The pipeline will then have long-run contracts to resell the gas with distributors in the markets. Clearly, short-run movements in costs and prices can only have a minor influence on gas movements 'short-run extraction is by its very nature a wasting, dwindling, activity, with relatively little management discretion' (Adelman, 1962, 10).

Having accepted that long-run parameters are to be considered, ideally it is the long-run *marginal* parameters which are needed. Unfortunately, there are several reasons why marginal data cannot be used in this study. First, 30 per cent of gas produced is found dissolved in oil, i.e., 30 per cent of gas output is found as a joint product in fixed proportions. For this portion of gas production there can be no marginal cost.[6]

Where gas is found by itself, marginal costs for gas production can theoretically exist. The problem is to measure them. What portion of a company's exploration costs should be attributed to gas production? Should the costs of a well which was sunk in an attempt to find gas but which struck oil be included in the exploration charges for a company's gas production? How should dry holes be treated? (Adelman, 1962, 31–82.) These issues show that it is difficult to conceive of individual firms having reliable estimates of their long-run marginal gas production costs.

It was stated above that the gas pipeline companies are not common carriers but they purchase the gas from the producer for resale. In 1954, the Federal Power Commission was given the authority to regulate the sales price for natural gas producers in inter-state commerce.[7] In 1960,

---

6 / Many methods have been advocated. One suggests dividing the costs on the relative BTU value of the fuels. Another divides the costs according to the relative prices of the fuels. Neither is sensible. The price of a commodity has no necessary connection to its cost. The intrinsic fuel–BTU value of a commodity in no way determines cost or price.

7 / *Phillips Petroleum Company* v. *Wisconsin,* 347 US 672 (1954).

the principle of regulating field prices by setting area-wide ceiling prices replaced the principle of pricing by the cost of service.[8] For each producing area under federal jurisdiction, the FPC sets the maximum price which producers can charge pipeline companies for gas.

The unit ceiling[9] prices set by the FPC become the unit costs of gas to inter-state pipeline companies. These prices are based on the concept of average cost.[10] The price which a producer can charge for a unit of gas is the same, no matter the volume sold. FPC area rate prices therefore represent the average and marginal costs of gas to an inter-state pipeline company.

Production costs in this model are assumed to be a linear and constant function of output.[11]

$$P(x_i) = p_i \sum_{j=1}^{n} x_{ij} \quad (\text{all } i, i = 1, ..., m). \tag{2.4}$$

$P(x_i)$ = total production costs for field i; p_i = the unit area rate price established by the FPC; x_{ij} = the flow of gas from field i to market j (MMCFD).

4 TRANSPORTATION COST

Figures 2.1 and 2.2 show the long-run total costs, average and marginal costs, for shipping alternative volumes of gas one mile. Each of the

---

8 / *Phillips Petroleum Company*, 29 FPC 338 (1960).

9 / In setting *ceiling* prices, the FPC is of the opinion that gas producers would act as monopolists without the FPC's intervention. Students of the industry are of the mind that price-setting is competitive but in the middle 1950s some *monopsonist* elements appears. See MacAvoy (1962); Adelman (1962).

10 / The FPC prices are for *pipeline quality* gas. The FPC calculates a national average cost of finding, developing, and producing gas. To this national average are added regional costs: quality improvement costs, royalty charges, plant processing costs and gathering costs. State production taxes and an allowance for a fair rate of return (percentage of investment) are also included. The inclusion and the method of measurement for each of these components can be questioned. Since these prices become the costs to the pipeline, it is irrelevant how they are arrived at.

11 / There are two further problems with using these FPC rates, which are discussed in the next chapter. First, the FPC rate applies only to *interstate* shipments. Intra-state flows may well be at different prices. The FPC has two rates, depending on the age of the contract. The question is which rate or what combination of rates suits this model best.

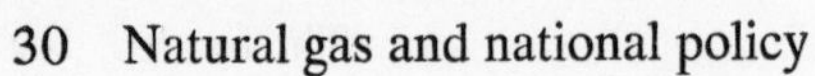

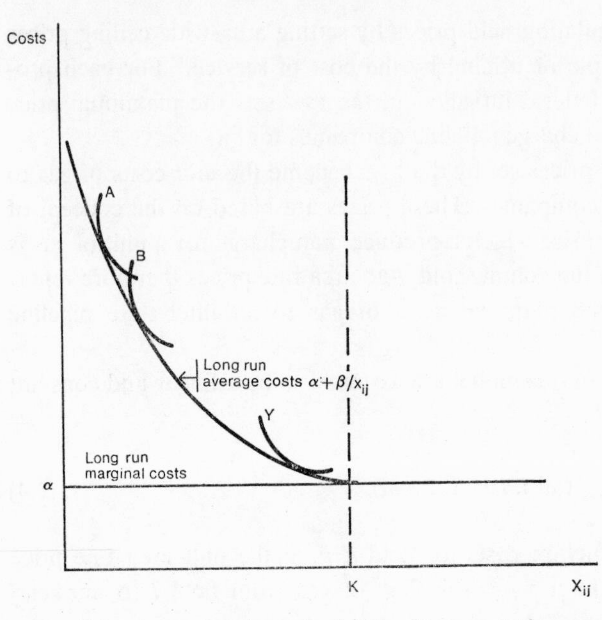

FIGURE 2.1 Average and marginal costs of transportation, per mile

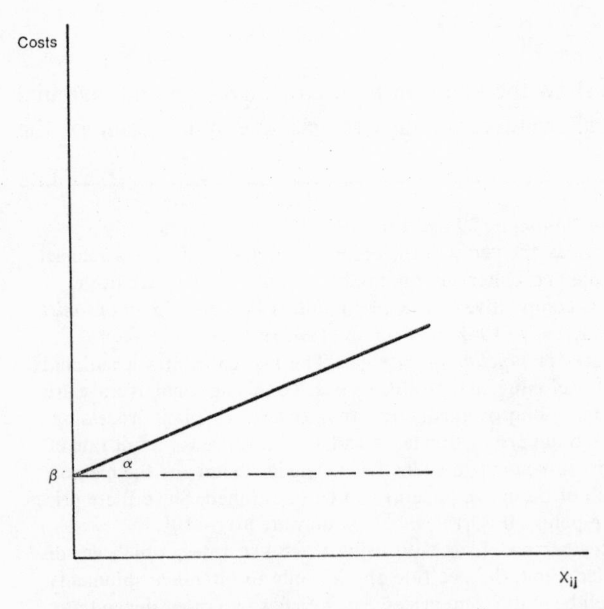

FIGURE 2.2 Total costs of transportation, per mile

curves is defined as the envelope for the various diameters of pipe which could be used. For example, the average cost curve in Figure 2.1 is drawn as the envelope for a number of short-run average cost curves, each for a different sized pipe. Curve *A* might be the average costs with a 6-inch pipe, *B* with an 8-inch pipe, *Y* with a 36-inch pipe. Each curve *A, B,* ..., *Y* assumes that the optimal choice has been made between pipe and horsepower. The point where each curve *A, B,* ..., *Y* touches the envelope determines a certain optimal load factor for both the individual curve and its envelope. Average costs per mile are assumed to fall until an absolute physical capacity (K) is met.[12] At capacity K, average costs equal the constant marginal cost. The transportation activity is not perfectly divisible – to ship one unit of gas requires that all the costs of constructing the minimum size pipeline be incurred. In addition, some of the fixed costs of building a pipeline are independent of the size of the line. For example, a single right of way for the pipe is necessary whether a 20-inch or 36-inch pipeline is being constructed. These two features – indivisibility and scale economies – suggest that the long-run curves are as depicted in Figures 2.1 and 2.2.

A total cost function (F_{ij}) which combines scale economies and indivisibilities is:

$$F_{ij} = ax_{ij} + b\lambda_{ij} \qquad (2.5)$$

($\lambda_{ij} = 1$, $x_{ij} > 0$; $\lambda_{ij} = 0$, $x_{ij} = 0$), where b represents the fixed costs necessary to ship from i to j and a the marginal costs for each unit shipped.[13] λ_{ij} shows that where the flow is zero, the fixed costs are zero, when the flow becomes positive, all the fixed costs are incurred.

The assumption of a constant marginal cost, a, is somewhat arbitrary, since the marginal costs of shipping gas may in fact fall with larger pipes if there are savings in labour or fuel costs. However, some simplifications are necessary to develop the model and this assumption is relatively innocuous.

---

12 / The physical capacity of a pipeline is not a determinate number like K. Capacity can be increased by adding horsepower to a station, a station to the line, or looping the line.

13 / The corresponding average cost function (f_{ij}) is:

$f_{ij} = F_{ij}/x_{ij} = a + b\,\lambda_{ij}/x_{ij}$.

The marginal cost is a:

$\partial F_{ij}/\partial x_{ij} = a$.

This per mile transportation function (F_{ij}) must be multiplied by the distance between the field i and market j (d_{ij}), to arrive at the total transportation costs of shipping from i to j.[14]

$$d_{ij}F_{ij} = ad_{ij}x_{ij} + b\lambda_{ij}d_{ij}. \tag{2.6}$$

The total costs of shipping between two points will be the same in (2.6) for all points the same distance apart. But it costs far more to build a pipeline and ship gas over a mountain or through muskeg than through prairie soil. Therefore, transportation costs are multiplied by a terrain factor (g_{ij}) which adjusts shipping costs for the topographic and geophysical characteristics of the land over which the flow must travel. This terrain factor was normalized at 1.0 for average prairie conditions and rises to 5.0 for the most difficult terrain.

The total adjusted transportation costs become:

$$d_{ij}g_{ij}F_{ij} = ad_{ij}g_{ij}x_{ij} + b\lambda_{ij}d_{ij}g_{ij}, \tag{2.7}$$

where b = the fixed costs per mile, incurred only if a positive flow is shipped; a = the marginal costs per mile; x_{ij} = the flow from field i to market j; d_{ij} = distance from field i to market j; g_{ij} = terrain factor for the route from i to j.

Were the model to attempt to minimize F_{ij}, severe computational problems would arise. The total cost function is non-convex between 0 and 1 units of flow. At a zero flow, no fixed costs are incurred. To send the smallest unit of gas, all the fixed costs, $bd_{ij}g_{ij}$, must be incurred. Non-convexities (indeterminancies) cannot be handled by linear programming.

The 'fixed charge' problem is in essence ignored, for the transportation costs which will appear in the objective function are only the long-run marginal costs = $ad_{ij}g_{ij}x_{ij}$.

The solution to a linear approximation to a fixed charge problem, where the fixed costs are ignored, will also yield an optimum to the true solution if all fixed charges are equal and the solution to the true problem is nondegenerate.[15] $bd_{ij}g_{ij}$ is not the same for all possible flows

14 / d_{ij} is calculated as the shortest distance between two points. If a lake is encountered, the distance is calculated around the obstacle.

15 / Hirsh and Dantzig (1954) were the first to derive this result. Nondegeneracy means that the number of positive flows will equal the number of independent constraints.

from fields to markets unless all fields are equally far apart from all markets and all terrain factors are equal.[16] This is certainly not the case.

How then can the omission of fixed costs tend to yield a correct solution? The answer lies along two intuitive grounds. If the problem were considering 20 fields and 200 markets, then ignoring fixed costs would see 220 separate flows from fields to markets.[17] However, a larger pipeline with lower average costs serving two or ten fields might reduce costs.[18] In this model, the level of *aggregation* is so high, 19 fields and 19 markets, that the size of most flows contemplated are only those which could be carried in the largest pipe (or multiples thereof). Savings could not be generated by letting one pipe serve a number of fields. Therefore a linear approximation in a highly aggregate model is not necessarily misleading.

Secondly, ignoring fixed costs will lead to an incorrect solution to the non-linear problem where the ratio of fixed to operating costs varies widely among different *i, j,* combinations. However, multiplying operating costs by g_{ij}, the terrain factor, ensures that there will be a close correspondence between operating and fixed costs.

5 THE MODEL

The objective function is:

$$\text{minimize } Z = \sum_{i=1}^{m} \sum_{j=1}^{n} (ad_{ij}g_{ij} + p_i)x_{ij}, \tag{2.8}$$

$$\text{subject to} \quad \sum_{i=1}^{m} x_{ij} \geq D_j \quad (j = 1, ..., n), \tag{2.9}$$

$$-\sum_{j=1}^{n} x_{ij} \geq -S_i \quad (i = 1, ..., m). \tag{2.10}$$ [19]

16 / I ignore the possibility that while d_{ij} and g_{ij} differ, their combinations are the same.

17 / There are 221 constraints, 200 market constraints, 20 field constraints, one surplus capacity constraint.

18 / If this were the case, then the solution is degenerate in the sense of n. 16.

19 / The convention of linear programming requires that inputs into a system be considered as non-positive. It is also easier for the development of the dual to have all the inequality constraints running in the same direction. If (2.10) is multiplied by -1, the more easily interpretable result of (2.3) follows, $\Sigma^n_{j=1} x_{ij} \leq S_i$. The total output of a field cannot exceed its capacity.

As I have argued in the previous section, the solution to this hypothetical free trade world will yield an approximation to the flows which would have appeared under a regime of free trade. In the next chapter, the important assumptions and drawbacks of the model are discussed. To look ahead, the plausible results obtained (chapter 4) and their insensitivity to large changes in the underlying data (chapter 7) reinforce this belief.

3
Data

1 SUPPLY

Supply points

Nineteen supply points were chosen for the USA and Canada (see Table 3.1). Canadian production is represented by four points, each point including an entire province's production. A finer breakdown of production was made for the USA. Of the fourteen American field nodes, seven represent specific fields.[1] The other seven include an entire state's production. One node was chosen to show the entry point of Mexican imports into the US.

It is difficult to represent an entire field or state's production by a single point. The point was chosen to be as near the geographic centre of production for the area, given the limitations of the terrain factor and distance calculating programs. For these programs, the co-ordinates of a node had to be even numbered degrees.

Capacity of fields

The daily capacity of a supply point was taken to be the actual production of the area represented by that point in 1966, divided by 365.[2]

1 / Data were available in the USA on a field basis for these seven regions. In Canada, all data were aggregated to provinces.

2 / The dual value (shadow value) of a field will therefore represent the marginal value of production.

TABLE 3.1
Supply points – co-ordinates and area included

Number	Name	Co-ordinates		Area included
		Latitude	Longitude	
1	British Columbia	120	60	Province
2	Alberta	114	56	"
3	Saskatchewan	106	52	"
4	Ontario	80	44	"
5	Mexico	100	28	Mexican imports into USA
6	Louisiana North	92	32	North of line drawn from eastern boundary with Mississippi
7	Louisiana South	92	30	South of Mississippi boundary plus offshore production
8	Texas Gulf	98	28	Texas railroad districts: 1, 2, 3, 4
9	Permian Basin	102	32	Texas railroad districts: 7C, 8
10	Mid-Continent	96	30	Texas railroad districts: 5, 6
11	Texas Panhandle	102	36	Texas railroad districts: 7B, 9, 10
12	San Juan Basin	108	36	New Mexico counties: San Juan, McKinley, Rio Argisa, and Sandoval
13	New Mexico Southeast	104	32	New Mexico counties: Hobbs, Chaves, and Eddy
14	Wyoming	110	42	State
15	Kansas	100	38	State
16	Oklahoma	96	36	State
17	Mississippi	90	32	State
18	California	120	36	State
19	West Virginia	80	40	State

For the purposes of this model, which is the reallocation of actual production to actual consumption, this is a good measure of capacity. The correspondence between this measure of capacity and any actual capacity is another issue.

The capacity of a well is a difficult concept to measure operationally. There are many possible definitions. The short-run physical capacity of a reservoir can be defined as the maximum rate which could be produced in one day without decreasing the potential supply (maximum efficient capacity (mec) for oil wells). This definition does not take into account the facilities necessary to produce and transport the gas to the market.

This physical reservoir measure of capacity may not be producible in reality. Production can be limited by the number of wells. (The number of wells is limited by regulatory authorities.) The actual capacity of the producing wells can be limited by the capacity of the gathering system. The gathering system may be able to transport quantities too large for the processing plant to handle. The processing plant may be forced to operate below its 'capacity' due to the inability of the trunk pipeline to handle more gas.

Each of these extensions is a possible measure of short-run capacity. If we consider possible measurements of long-run capacity, the permutations and combinations of adding some unit to each of the above possible short-run bottlenecks could be the subject of a paper itself. Therefore actual production is the measure of supply used in the estimations presented in the next chapter. However, as was pointed out in the introduction, the prevention of trade by the authorities must affect the pattern of production. An alternative measure of supply is given in column 2 of Table 3.2. The average reserve to production ratio for all producing areas in North America was calculated to be 16.85 for 1966. Some production points, namely Alberta, British Columbia, California, and Kansas were well above the average (Alberta's ratio was 38.4), while some were far below – Northern Louisiana (8.4), New Mexico SE (10.8), West Virginia (11.8), and the Texas gulf (11.8). Column 2 is derived by assuming that all regions could produce at the average reserve/production ratio for North America. The estimate of supply in areas such as Alberta rises while it falls for fields such as New Mexico SE. The total amount of production is the same in columns 1 and 2 of Table 3.2.

This is a rather plausible change to make. The effect of both the policy to maintain an inventory of gas in the ground in Canada for future Canadian demand, and the policy restricting trade, tends to increase the reserve to production ratio in western Canada. This second estimate of supply can then be used to see whether the results obtained from the simple model which reallocates actual production to actual demand are biased because of the assumption that supply equals actual production. These results are given in chapter 7.

2 DEMAND

Nineteen demand points were used in this model, nine of which were in Canada (Table 3.3). The demand areas are very gross, ranging from

TABLE 3.2
Capacity, 1966

	Area	Daily marketed production 1966* MMFCD 14.73 psi	Alternative supply reserve/production = 16.85 MMFCD 14.73 psi
1	British Columbia	441	920
2	Alberta	2441	5561
3	Saskatchewan	103	112
4	Ontario	43	31
5	Mexico	150	150†
6	Louisiana N	1667	835
7	Louisiana S	12,255	12,656
8	Texas Gulf	10,662	7444
9	Permian	3550	3340
10	Mid-Continent	1790	1744
11	Texas Panhandle	3250	2766
12	San Juan	1397	1564
13	New Mexico SE	1338	854
14	Wyoming	667	612
15	Kansas	2322	3186
16	Oklahoma	3702	3309
17	Mississippi	429	321
18	California	1889	2871
19	West Virginia	580	405
		48676	48621

* / Net production: i.e., gross production less volumes re-injected.
† / No data available to change the estimate of Mexican production.
SOURCES: Daily marketed production, (*a*) USA and Mexico: Bureau of Mines, *Natural Gas Production and Consumption* (1966); (*b*) Canada: Canadian Petroleum Association, *1966 Statistical Yearbook*; reserves: American Gas Association, *Reserves*, 1967

entire provinces in Canada to groups of states in the USA. The breakdown in the USA was identical to that of the Future Requirements Committee[3] and was used to take advantage of that Committee's demand projections.

It is obviously difficult to pick a single point as representative of each of these large regions. Weighting each set of co-ordinates in each region by the consumption of gas at that point would yield an average consuming point. The average distance travelled by each MCF of gas through each region would represent yet another possible point. A third possibility would be to pick points which maximized the distance that gas had to travel. The method actually used was a combination of all three, leaning heavily on the third. In region USA 1, the largest

3 / See *Oil and Gas Journal* (July 17, 1967), 44.

TABLE 3.3
Demand points – co-ordinates and area included

	Point	Co-ordinates		Area included
		Latitude	Longitude	
1	British Columbia	122	50	Province
2	Alberta	114	51	”
3	Saskatchewan	106	52	”
4	Manitoba	90	50	”
5	Ontario Northwest	84	48	Port Arthur, Sudbury, Orillia
6	Ontario Southwest	81	43	Windsor and environs
7	Ontario–Toronto	79	44	Toronto
8	Ontario East	76	45	Ottawa, Belleville, Kingston, Maitland
9	Quebec–Montreal	74	45	
10	US1	71	43	Maine, Vermont, New Hampshire, Massachusetts, Rhode Island, Connecticut
11	US2	74	40	New York, Pennsylvania, Ohio, Maryland, Delaware, New Jersey, DC, Virginia, West Virginia, Kentucky
12	US3	84	34	Tennessee, North Carolina, South Carolina, Alabama, Georgia, Florida
13	US4	88	41	Wisconsin, Michigan, Illinois, Indiana
14	US5	96	44	North Dakota, South Dakota, Minnesota, Iowa, Nebraska
15	US6	95	37	Kansas, Missouri, Oklahoma
16	US7	94	33	Texas, Arkansas, Mississippi, Louisiana
17	US8	108	46	Montana, Wyoming, Utah, Colorado
18	US9	124	38	New Mexico, Arizona, Nevada, California
19	US10	124	46	Washington, Oregon, Idaho

consuming point in the area is Boston,[4] with little consumption north of that city; therefore the consuming node was chosen to be just south of Boston. In region USA 9, San Francisco was the choice for the demand point, even though large quantities of gas are consumed in New Mexico. Some measure of reality was attempted by choosing demand points to approximate the ultimate end nodes of existing gas systems.

Ontario was broken down into four regions. To best represent the realities which led to the construction of the trans-Canada pipeline,

4 / Remember that the co-ordinates of each point had to be in even numbered degrees.

TABLE 3.4
Demand

		1966 consumption* MMCFD 14.73 psi	Future demand 1975	Future demand 1980
1	British Columbia	219	444	545
2	Alberta	572	797	967
3	Saskatchewan	176	350	430
4	Manitoba	103	244	304
5	Ontario Northwest	146	308	395
6	Ontario Southwest	187	395	510
7	Ontario–Toronto	243	515	660
8	Ontario Southeast	89	190	240
9	Quebec–Montreal	89	182	260
10	US1	524	820	970
11	US2	8016	11,240	12,800
12	US3	3025	4575	5310
13	US4	5884	8235	9520
14	US5	1978	2830	3665
15	US6	3846	4610	4695
16	US7	15,324	24,585	27,900
17	US8	979	1765	2000
18	US9	6396	10,030	10,780
19	US10	604	1295	1500
		48,400	73,410	83,451

* / Including transmission losses.
SOURCES: (*a*) Canada except Ontario: Canadian Gas Association, *Canadian Gas Facts 1967*; (*b*) Ontario: the total Ontario demand, obtained from *Canadian Gas Facts*, was broken down to regions by a percentage of total calculated from Trans Canada Pipelines, *Application October 1967* to the National Energy Board; (*c*) USA: Bureau of Mines, *Natural Gas Production and Consumption, 1966*.
NOTE: For areas which are not, in this model, considered as supply points, their production is *deducted* from their demand.
(*d*) Future demand, USA: *Oil and Gas Journal*, July 17, 1967; Canada: National Energy Board, *Energy Supply and Demand Forecast, 1965–1985*, Ottawa, 1967.

consumption points were introduced for actual end points of the existing pipeline.

The first column of Table 3.4 presents the actual consumption data for 1966. The next two columns give estimates of the demand in these regions at two future dates – 1975 and 1980. Solutions to the model are obtained using these alternative demand estimates. Supply must also be increased so as to meet these greater demand estimates. For both the 1975 and the 1980 estimates of total demand, a reserve to production ratio is calculated using 1966 total reserves. Each field is assumed to produce at this ratio.

These exercises are purely hypothetical, for by 1975 reserves will have changed. The purpose is not, however, to exactly portray the future, but to see if changes in the underlying data significantly affect the basic results of the simple model. The results using these estimates of demand and supply are given in chapter 7.

3 FIELD PRICES

Inter-state sales

As was stated in the chapter which developed the model, the area rate prices established by the Federal Power Commission will be used to represent the production costs in the fields.

The FPC, however, regulates only inter-state flows of gas. This model applies the area rate price to all flows of gas, inter-state and intra-state. Two possible errors result. Inter-state sales may be made at less than the ceiling price. Intra-state shipments may be made at different prices.

It is difficult to evaluate the first possible error. Producers have tried to limit regulation. These efforts at ending or changing price controls have a cost. Without regulation, the prevailing price would therefore be expected to be higher. Table 3.5 gives the annual changes of the average field price on all inter-state sales by producers to pipeline companies. The impact of the 1960 decision to establish ceiling prices can be seen at the beginning of 1962 by the slowdown of the rate of increase of gas prices. It is assumed that the area rate prices are binding constraints on inter-state gas shipments.

The Commission has established three types of area prices: area price ceilings, in-line rates, and initial guideline rates. *Area price ceilings* are determined after the Commission and relevant producers present cost and operating statistics before an examiner. *In-line rates* are rates established by the Commission while the ceiling price is being determined. No consideration is given to cost data. The Commission determines acceptable prices on the basis of the general price level in the area concerned. *Guideline rates* are the prices which the Commission establishes for new contracts in areas where no formal rate hearing is being undertaken.

As of December 31, 1966, only one area rate ceiling was in effect. The initial examiner had established ceiling prices for the Permian

TABLE 3.5
Field prices of natural gas charged by producers to pipeline companies, 1953–65

	Price (mid-year) ¢MCF, 14.65 psi	Change from previous year (%)
1953	9.1	
1954	10.0	9.9
1955	10.7	7.0
1956	11.2	4.7
1957	12.0	7.2
1958	13.0	8.3
1959	14.4	10.8
1960	15.6	8.3
1961	16.5	5.8
1962	16.8	1.8
1963	16.9	0.6
1964	16.7	(1.2)
1965	16.8	0.5

SOURCE: Interpolation of graph, Federal Power Commission, *1965 Report*, 106.

Basin in 1964.[5] Four other areas had proceedings under way: South Louisiana, Hugoton-Andarko, Texas Gulf Coast, and other Southwest (Arkansas, Mississippi, four counties in Alabama, northern Louisiana, Texas Railroad Commission districts 5, 6 and 9, and most of Oklahoma).

These holding tactics of the Commission do not have to be followed by the producers. Producers can negotiate higher prices with the pipeline companies in the hope of raising the FPC ceiling at some point in the lengthy court proceedings. If this were a representation of producer behaviour, using these FPC prices would lead one to underprice gas.

It is unlikely that these higher prices do exist. In the in-line and guideline rates the Commission sets the rate according to the average price level in that area. It would be difficult for a producer to negotiate a competitive price with a buyer which is higher than the going price for other gas reserves in that area. In the case of area price ceilings which are based on cost, not price data, an individual producer may be able to negotiate a higher price. However, the volume of repayments[6] with interest ordered by the Commission when a producer is charging

5 / Early in 1968, the Supreme Court reaffirmed the right of the Federal Power Commission to regulate area rates in the Permian Basin case.

6 / Also, when the Commission lowers an in-line rate in an area ceiling proceeding, producers are liable for refunds plus interest. This is another argument why producers will not negotiate prices above the in-line rate.

a price higher than the final area ceiling price leads one to believe that producers do follow the Commission set rates.

It has been shown that the various Federal Power Commission regulated area prices are a true constraint; producers try to remove them, yet are reluctant to go above them. Since they are below the level which would be set by market forces, it is unlikely that producers would commit reserves at prices *lower* than these rates. The FPC rates therefore must be the prices in existence to pipeline companies. These prices represent the average cost of gas to inter-state pipeline companies.

Intra-state sales

Shipments of intra-state gas are a large proportion of total gas flows. Table 3.6 shows the domestic sales to inter-state pipeline companies as a ratio of total marketed production. Inter-state sales had represented an increasing proportion of marketed production up to 1960, the year the area rate price was introduced. With the limited evidence available, it is not possible to deduce whether in fact the introduction of the area rates did dampen the growth of inter-state sales or whether other factors such as differential growth rates were the cause.[7] If the competitively determined intra-state price was higher than the inter-state area rate price, a large swing to intra-state flows would appear. Intra-state flows are not in fact increasing at a faster rate than inter-state flows. The price of intra-state flows cannot be far different from the area rate price established by the regulatory authorities for inter-state flows.[8]

Old gas contracts and flowing gas contracts

The Federal Power Commission has established two sets of area prices depending on the age of the contract. The area prices for contracts

7 / Inter-state pipelines ship mainly to the northeast, mid-west, and far west. If the industrial growth rate of the southwest increased relative to these other areas, *ceteris paribus*, the percentage of inter-state flows would decrease.

8 / For the period 1953–8, including three years before and three years after the initial Phillips decision, Gerwig (1962) finds that prices on inter-state sales were significantly higher than on intra-state sales. He attributes this to the increased cost due to administrative delays and regulatory risks of inter-state sales. He does state, however, that definitive price ceilings would result in a pattern of costs, and therefore prices different from those observed in the uncertain period following the initial 1954 decision.

TABLE 3.6
Inter-state sales as a percentage of all gas sales, 1953–66

Year (1)	Sales to inter-state pipelines less Canadian imports (MMCF) (2)	Marketed production (MMCF) (3)	Percentage (2)/(3) (4)
1953	4713	8397	56.1
1954	5134	8743	58.7
1955	5527	9405	58.8
1956	6145	10,082	61.0
1957	6518	10,680	61.0
1958	6857	11,030	62.2
1959	7520	12,046	62.4
1960	8009	12,771	62.7
1961	8001	13,254	60.4
1962	8422	13,877	60.7
1963	9656	17,477	62.1
1964	9670	15,462	62.5
1965	9910	16,040	61.8
1966	10701	17,207	62.1

SOURCE: Sales to interstate pipelines: American Gas Association, *Gas Facts 1968*, 34; Canadian imports: National Energy Board, *Report*, 1966, appendix VIII; marketed production: American Gas Association, *Gas facts 1968*, 28.

signed after January 1, 1961 (flowing gas contracts) are at least as great as and generally higher than the rates established for contracts in existence before 1961 (old gas contracts). To confuse the issue further, any one area may have in-line rates and initial guideline prices in effect for both old and new contracts at the same time. This occurs when the Commission begins a rate proceeding in an area where guideline prices previously existed. Columns 2 and 3 of Table 3.7 show two sets of prices which were used in the model estimates, one for new contracts and one for old contracts. If an in-line rate existed for an area, it was used rather than the guideline rate. The appropriate question is which price set will fit this model best. Which of these price sets will most closely parallel the price parameters that operators considered when making the decisions which resulted in the actual 1966 flows? If the purpose of the model is to determine the change in flows in 1966 when restraints on trade are hypothetically lifted, then it is the new contract prices which are relevant. It is these prices which determine the flows of new supplies. On the other hand, it could be argued that the old contract prices are relevant since the larger share of gas flowing in 1966 was committed before 1961. The 1966 flows were not, however, determined by either of these two price sets alone. Both sets of contracts were in existence. The problem then becomes how to combine the two sets of price estimates in

TABLE 3.7
Federal Power Commission area rate prices, new contracts, old contracts, combined prices, 1966

Area (1)	FPC area rate price (/MCF) (14.73 psi)		
	New contracts (post-1961) (2)	Old contracts (pre-1961) (3)	Combined (4)
Louisiana North (5)	16.67	14.08	15.61
Louisiana South (6)*	19.39	13.73	16.12
Texas Gulf (6)*	15.83	14.08	14.54
Texas Permian (7)	16.59	14.58	14.87
Texas Midcontinent (5)*	15.00	14.16	14.37
Texas Panhandle (5)*	16.34	11.69	13.71
San Juan (5)	12.74	12.74	12.74
New Mexico Southeast (7)	15.58	13.57	14.07
Wyoming (6)	15.08	12.74	13.79
Kansas (5)	16.89	11.06	11.06†
Oklahoma (6)	17.09	11.06	14.16
Mississippi (6)	20.71	14.08	16.21
West Virginia (5)	26.91	24.03	25.62

NOTES: (5) Initial guideline rate; (6) in-line rate; (7) Area rate price ceiling, per docket no. AR 61–1, Proceedings (opinion no. 408).

* / These areas in this model are a combination of FPC areas. FPC prices are weighted by 1966 production to give an average rate

† / 100 per cent of 1966 flow estimated as coming from reserves established prior to 1961.

Data from Federal Power Commission files.

some meaningful way to arrive at an average price for 1966 production.

As an approximation, one can measure how important reserves established before 1961 are in 1966 production. Assume for the moment that the 1960 production-to-reserve ratio was maintained for these 1960 reserves for each year up to and including 1966. The production available in 1961 from reserves established prior to 1961 was:

$$P_{61} = 0.05\ (R_{60}), \tag{3.1}$$

$$R_{61} = R_{60} - 0.05\ (R_{60}). \tag{3.2}$$

P_{61} = production in year 1961 from reserves established prior to 1961; R_{60} = reserves at end of 1960; R_{61} = reserves available at end of 1961 from reserves established prior to 1961.

Likewise for 1962:

$$P_{62} = 0.05\ (R_{61}), \tag{3.3}$$

$$= 0.05\ (R_{60} - P_{61}), \tag{3.4}$$

$$= 0.05\ (R_{60} - 0.05\ (R_{60})), \tag{3.5}$$

where P_{62} = production in year 1962 from reserves established prior to 1961; and for 1966

$$P_{66} = (0.05)R_{60} - 5(0.05)^2R_{60} + 10(0.05)^3R_{60} - 10(0.05)^4R_{60} + 5(0.05)^5R_{60} - (0.05)^6R_{60}. \qquad (3.6)$$

Substituting into expression (3.6) the value for 1960 reserves, production in 1966 from the reserves established prior to 1961 is 10060 MMCF, or 57 per cent of actual 1966 production.

The two price sets, established by the Federal Power Commission for flowing and for old gas, could be combined using this value.

To combine the data in such a manner involves many assumptions. First, it is not clear that the maintenance of a constant reserve production ratio represents rational producer behaviour. If rational behaviour was to maintain the same absolute amount of production in each year as was established in 1960 from the reserves available in that year, 75 per cent of actual 1966 production is from pre-1961 reserves. It is also an unwarranted assumption to believe that all fields either have the same pattern of reserve acquisition or maintain the same reserve production ratio. Therefore, for each producing area, the production from reserves established prior to 1961 was calculated using equation (3.6).

Column 4 of Table 3.7 gives the price set which is a weighted combination of the FPC old and flowing gas price sets, the weights for each field depending on the percentage of production estimated as flowing from reserves established both before and after 1961.

All three price sets are used in the model. Since the prices for a field vary greatly between its old and its new price, using these different price sets will test the sensitivity of the more important results to the price data used.

Non-regulated areas

There are producing areas in our model which are not regulated by the Federal Power Commission: Canada, Mexico, and the purely intrastate production of California. The prices used for these areas are presented below in Table 3.8.

For the American fields, the average purchase price to inter-state pipeline companies has been used. When available, a similar price was used

TABLE 3.8
1966 area prices, unregulated areas

	14.73 psi, American dollars
British Columbia	10.60
Alberta	13.00
Saskatchewan	16.55
Ontario	38.00
Mexico	17.16
California	30.60

SOURCES: British Columbia: average purchase price paid by Westcoast Transmission (including gathering and processing costs), including two years' price escalation (0.25/year), weighted by contract size; Alberta: average purchase price paid by the Trans Canada Pipeline Company for saleable gas (including gathering and processing costs), including two years' price escalation (0.25/year), weighted by contract size; Saskatchewan: contract price paid by the Trans Canada Pipeline Company at Steelman processing plant, plus two years' escalation (0.25/year); Ontario: average well head price; private communication with the Ontario Energy Board; Mexico: *Oil and Gas Journal*, April 4, 1966, 147; California: average value for field use. *Natural Gas Consumption and Production*, 1966, 5.

for these other areas. Otherwise, the average price of field use (average price for use within the area – re-injection to increase oil flow) is taken. The first column of Table 3.8 gives these prices. In chapter 7, these production price estimates are revised to determine how significant the prices are in determining the models' basic results.

4 THE TRANSPORTATION COST FUNCTION

In the previous chapter, an average cost function was derived of the following form:

$f_{ij} = a + b/x_{ij}$ (see p. 31, n. 13).

This function was fitted to average cost data given in a paper by Dr C.L. Dunn.[9] The following equation fitted this data best:[10]

$$f_{ij} = 1.1 + 140/x_{ij}. \quad (3.7)$$

f_{ij} = cents/MCFD/100 miles; x_{ij} = flow from i to j in MCFD.

9 / 'The economics of Gas Transmission,' INGAA Conference (Oct. 1959).
10 / $R_2 = 0.89$, significant at 1 per cent level.

For various flows, the average cost of transmission is as follows[11] (plotted in Figure 3.1): flow (MMCF): 100, 200, 600, 1000, 2000; cost ¢/MCF/100 miles: 2.5, 1.8, 1.35, 1.25, 1.17.

From equation 3.7, the fixed and marginal costs of transportation per MCF per mile are 1.40 cents and 0.011 cents respectively. The value of *a* in these tests is then 0.011 cents per MCF per mile.

To test the sensitivity of the important results to these estimates of transportation costs, several other estimates both higher and lower are also used. Chapter 7 develops these findings.

5 ASSUMPTIONS LIMITATIONS OF THE MODEL FORMAT

In this section the various assumptions will be reiterated. What the model can and cannot do will be clearly stated. The purpose of this exercise is

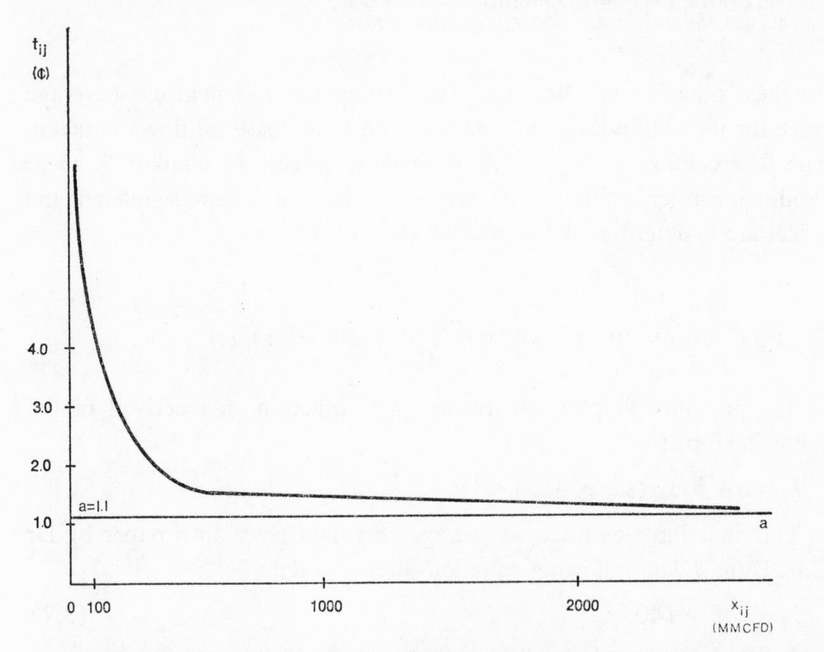

FIGURE 3.1 Average and marginal costs of transportation (t_{ij}), ¢/MCF/100 miles

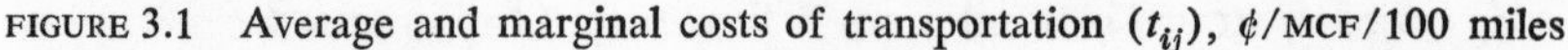

11 / Adelman (1962), 49, reckons that the cost per 1000 miles on the Gulf Coast-New York route would be 13.9 cents per MCF. MacAvoy (1962) finds lower

to examine the effects of a constraint which prevents shipments of gas between the United States and Canada. A linear programming model incorporating actual 1966 cost and sales data is used to find the flow pattern which will minimize the final costs at the market.

Since 1966 sales and costs are used, the model will not find the *optimum optimorum* flow solution. The data used are the result of the past workings of a system where the border constraint has been operative. As a result, the costs and sales data for 1966 contain elements related to the impact of the border in 1965, 1964, 1963 ... To find the *optimum optimorum* flow pattern for 1966 would involve a model which arrived at year by year solutions without a border for the North American industry from the date of inception of that industry. The model presented in this book is a static one-year solution. It will yield a second best estimate of the costs of the border, an estimate that will be lower than the true value of these costs. This model finds a solution as if the industry were allowed in 1966 to establish a flow pattern in the absence of a border but constrained to meet the demand, supply, and cost restrictions which would have resulted from that border's existence.

Assumptions

1 *Profit maximization* (cost minimization). This is the behavioural assumption.

2 *Linear and homogeneous costs of production.* It costs the same to produce the first unit of gas and the millionth unit. Field prices are exogenously determined.

3 *Demand is perfectly inelastic.* The sum of all sellers (the industry) faces a perfectly inelastic demand curve. The quantity demanded does not change with a change in price. The free trade solution, it is hypothesized, will yield a lower total cost solution than the actual system. Some points in the system will therefore be supplied in the free trade solution at a lower price than they are now being supplied at. Assuming that gas is a superior good, the demand for it will *rise* at these points. This increase in demand is not recognized by the model. The cross elasticity of demand between gas and other fuels is likewise ignored.

---

average transportion costs. His estimates range from 0.60 cents (a 36-inch pipe and an average daily flow of 1265 MCF) to 1.01 cents (a 20-inch pipe and an average daily flow of 175 MCF) per 100 miles.

4 *Total supply has an upper limit. This supply is price inelastic.* The amount of excess capacity in the system is determined solely by the relation of total supply to total demand. The total available capacity of the system is 48676 MMCFD. The total demand is 48400 MMCFD 276 MMCFD will be excess capacity since the second constraint states that demand must be met in each market, regardless of price.

Supply does not change in total[12] to a change in price at demand points. Total supply is perfectly price inelastic.

If total demand and total supply are both perfectly inelastic, it would appear that price is indeterminate. Price, in this model, is established in each individual market as the lowest price which will cover the costs of the marginal supplier considering that the totality of all markets must be supplied at lowest cost. See Figure 3.2.

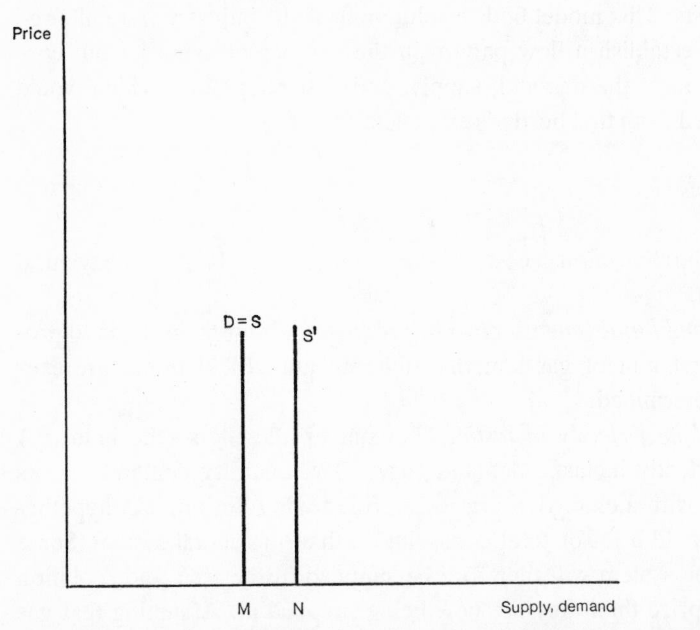

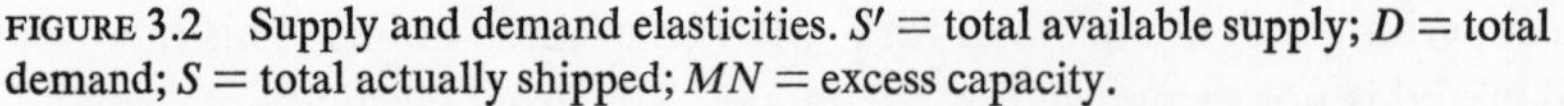
FIGURE 3.2 Supply and demand elasticities. S' = total available supply; D = total demand; S = total actually shipped; MN = excess capacity.

---

12 / The quantity supplied by an individual field does change in response to changes in *relative prices*. The excess capacity will be borne by different fields as prices change relatively. No change in absolute prices, relative prices remaining constant, can affect the quantity supplied by individual fields.

Limitations

The system is formulated as a single-period static model. As such it has certain limitations.

1 *There is no past.* No past is analogous to perfect foresight. No existing pipelines are assumed. No errors exist. Excess capacity in the transportation system does not exist.

2 *There is no future.* Transport investment decisions are not made. The interesting and important question of when to build a pipeline is not asked. The model prices transportation by looking only at long-run marginal costs. As has been discussed this is not a major drawback for determining *where* the optimal distribution system should be rather than *when* it should be built. Alternatively, the study could have proceeded by considering the investment cost of pipelines joining all pairs of nodes. The objective would then be the minimization of the sum of investment costs, pipeline operating costs, and production costs. No attempts were made to solve this non-linear integer programming problem.

4

Results of the free trade model, 1966 data, and development of the constrained model

1 FREE TRADE SOLUTIONS

The linear programming free trade model as given by equations 2.8 to 2.10 in chapter 2, was solved using the basic data for 1966. In this section, the actual production and demand data for 1966 are used as are the estimated cost and price data given in chapter 3, Tables 3.2 to 3.8 and equation 3.7.

Three alternative estimates of production costs for the USA were given in chapter 3. All three estimates were used in deriving solutions to the free trade model.[1] These solutions are given in Tables 4.1 and 4.2, the former being the solution for both the old or combined price sets, the latter using the new price series. Although, relative prices differ markedly between the three price series (50 per cent between old and new contract prices), the solutions are nearly identical. The only difference is that some excess capacity shifts from field 9 to field 7 when the new price series is used as the estimate of production costs in the USA. The free trade model is insensitive to substantial changes in relative production costs.

---

1 / Each solution involved 38 independent equations (in total, there are 39 equations – 19 supply equations, 19 demand equations, and one surplus capacity equation; of these one is dependent – can be expressed as a linear combination of the remaining 38), and 399 variables. Note that only 37 flows are reported in the free trade solutions. There is, in addition, a small positive flow from Saskatchewan to Manitoba. Increasing the estimate of capacity at Saskatchewan, with or without increasing demand at Manitoba, increases this small flow. No changes in dual variables occurred when supply was so increased.

TABLE 4.1
Primal flows: free trade model, old contract prices or combined prices, MMCFD

From \ To		BC 1	Alta 2	Sask. 3	Man. 4	Ont. NW 5	Ont. SW 6	Ont.–Tor. 7	Ont. SE 8	Quebec 9	US1 10	US2 11	US3 12	US4 13	US5 14	US6 15	US7 16	US8 17	US9 18	US10 19	Slack
British Columbia	1	219																		222	
Alberta	2		572	176														979	331	382	
Saskatchewan	3				103																
Ontario	4																				43
Mexico	5																				150
Louisiana North	6						187		89					1391							
Louisiana South	7											7960	2596	1189			510				
Texas gulf	8																10,662				
Texas Permian	9							243		89							2362		773		83
Texas Midcontinent	10																1790				
Texas Panhandle	11															3250					
San Juan	12																		1397		
New Mexico SE	13																		1338		
Wyoming	14																		667		
Kansas	15					146								198	1973						
Oklahoma	16													3106		596					
Mississippi	17												429								
California	18																		1889		
West Virginia	19										524	56									

TABLE 4.2
Primal flows: free trade model, new contract prices, MMCFD

From \ To		BC 1	Alta 2	Sask. 3	Man. 4	Ont. NW 5	Ont. SW 6	Ont.–Tor. 7	Ont.SE 8	Quebec 9	US1 10	US2 11	US3 12	US4 13	US5 14	US6 15	US7 16	US8 17	US9 18	US10 19	Slack
British Columbia	1	219																		222	
Alberta	2		572	176														979	331	382	
Saskatchewan	3				103																
Ontario	4																				43
Mexico	5																				150
Louisiana North	6						187		89					1391							
Louisiana South	7											7960	2596	1189			427				83
Texas Gulf	8																10,662				
Texas Permian	9							243		89							2445		773		
Texas Midcontinent	10																1790				
Texas Panhandle	11															3250					
San Juan	12																		1397		
New Mexico SE	13																		1338		
Wyoming	14																		667		
Kansas	15					146								198	1978						
Oklahoma	16													3106		596					
Mississippi	17												429								
California	18																		1889		
West Virginia	19										524	56									

In the real world of 1966, North America could be divided into two rather distinct trading areas, the division being the 49th parallel. Except for the supplies from western Canada to the USA and the small shipments from the USA to eastern Canada, two distinct non-trading areas existed, one in the north and one in the south.

The most significant aspect of Tables 4.1 and 4.2 is the complete absence of any transcontinental east-west gas flows. In particular, the shipments from Alberta to eastern Canada which are a dominant feature of life do not appear in these hypothetical solutions to a free trade model.

In this hypothetical free trade world, two distinct non-trading areas can be discerned. These however are not divided into north-south areas as they are today but into western and eastern regions. The western zone contains seven markets (British Columbia, Alberta, Saskatchewan, Manitoba, US 8 (Montana, Utah, Wyoming, and Colorado), US 9 (California), and US 10 (Pacific Northwest)). The western zone also includes seven fields (British Columbia, Alberta, Saskatchewan, Wyoming, San Juan, New Mexico southwest, and California). The eastern trading zone contains the remaining twelve markets and twelve fields.

In this free trade model, no trade flows appear between these two areas. The natural division for trade in natural gas would not appear to be the division between the USA and Canada, but one between eastern and western North America.[2]

2 A COMPARISON BETWEEN THE FREE TRADE AND THE ACTUAL FLOWS

Table 4.3 compares the actual 1966 flows from broad producing areas with the flows as estimated in the free trade solutions (and in the constrained solutions which will shortly be derived).

The agreement between the actual flows and the flows as estimated in the free trade model is quite close. The actual flows are determined by the capacities of the existing pipelines which were built under conditions of imperfect knowledge and uncertain foresight. The free trade flow model

2 / The foresight of Canadian politicians is proved to be correct. Without constraints on imports and exports, trade flows in natural gas would be on north-south lines.

TABLE 4.3
Comparison of 1966 actual and estimated flows, MMCFD

Shipment		Flows		
			Estimated	
From	To	Actual (approximate)	Free trade	Constrained trade
Canada	US 8, 9, 10	1060	1915	1283
Western Canada	Eastern Canada	632	—	632
USA	Eastern Canada	122	754	122
New Mexico	California	2860	2735	2735
US West South Central (Louisiana, Oklahoma, Texas)	US 1, 2	7300	7960	7960
–	US 4, 5	5020	5686	5480
–	US 7	15300	15320	15320
– (and Kansas)	US 6	3470	3846	3846
Louisiana and Mississippi	US 7	2800	3025	2800
West Virginia	US 2	560	524	524

SOURCE: Actual flows: USA: in 1966 the Bureau of Mines discontinued the classification of the imports into a region by producing area. The figures here are for 1965 (Bureau of Mines, *Consumption and Production*) increased by that region's consumption growth factor between 1965 and 1966; *Canada*: National Energy Board, *Report*, 1966.

does not assume a capacity restriction on any arc. The grossness of the model and its seemingly restrictive assumptions would have, at first glance, destroyed its applicability. The similarity between the two flows is interesting. The model appears to have escaped its limitations.

The shipments from American fields to American markets in the hypothetical solution most closely parallel the real world in both direction and size. New Mexico does ship nearly its entire production to California. The lines from the Louisiana and Texas fields to eastern markets parallel real lines. Chicago's demand is met with similar supplies in the actual and hypothetical worlds.

The large relative differences between the real and free trade worlds involve shipments across the border. In 1966, Canada exported 46.0 per cent of its marketed production. In the free trade solutions, 68.0 per cent of Canadian marketed production was exported. Canadian exports to the USA represented, in 1966, 2.5 per cent of that country's demand. In the solution to the model 5.6 per cent of American demand is met by Canadian producers. In 1966, imports into Canada from the USA represented 6.7 per cent of Canadian consumption. In the free trade solutions, 41.3 per cent of Canadian demand is met by American suppliers.

The hypotheses developed in the introductory chapters are observed in these solutions. The actual flow pattern for 1966 is close to the free trade pattern except for the shipments which are subject to the approval of the regulatory authorities.

3 A CANADIAN EAST-WEST PIPELINE – THE CONSTRAINED SOLUTIONS

The major differences between these solutions to the basic model with free trade and the actual flows for 1966 are the international flows between Canada and the USA.

To simulate more closely the actual movement of natural gas within North America, the free trade model is altered by substituting a set of constraints which incorporate the political realities of trade. Under present Canadian policy, western Canada must ship through to 'meet demand' at eastern Canadian markets. The surplus of capacity in the Canadian west above eastern Canadian demand is then exportable.

The *constrained* model can be represented by:

$$\text{minimize } z = \sum_{i=1}^{m} \sum_{j=1}^{n} [ad_{ij}g_{ij} + p_i]x_{ij}, \tag{4.1}$$

$$\text{subject to } x_{kj} \geq D_j \quad (j = 1, \dots h) \tag{4.2}$$

$$\sum_{i=1}^{m} x_{ij} \geq D_j \quad (j = h + 1, \dots, n) \tag{4.3}$$

$$-\sum_{j=1}^{n} x_{ij} \geq -S_i \quad (i = 1, \dots, n). \tag{4.4}$$

The objective function is identical in both the free trade and constrained models, as is the last constraint (2.10 and 4.4), that no field ship more than its capacity. Constraint 4.2 states that the shipments from the kth field (Alberta) to the markets 1, ..., h (eastern Canada) must at least be as great as the actual shipments in 1966.[3] This constraint set is analogous to the form of regulation. Alberta is constrained to meet demand in

---

3 / Imports into Canada were not zero. 121 MMCFD were imported into Ontario Southwest. Alberta was actually forced to meet 66 MCFD in this market (187-121).

eastern Canada.[4] Constraint set 4.3 states that the sum of flows into any one of the remaining markets (outside eastern Canada) from all the sources to that market must at least meet market demand.

The model as given by equation set 4.1 through 4.5 was re-run for each of the three price series. The flow solutions to these three runs are called the *constrained* solutions (as opposed to the free trade solutions where Alberta is not 'constrained' to meet eastern Canadian demand). These solutions are shown in Tables 4.4 and 4.5, Table 4.4 giving the flows when the old contract price or combined price series are used as the estimates of production cost in the USA while Table 4.5 uses new contract prices. There is a close correspondence between the actual flows for 1966 and the flows of the constrained solution (Table 4.3).

Table 4.6 gives the differences in flows between the constrained and the free trade solutions. These differences are the same, no matter which of the three alternative estimates is used for production costs in American fields.

The major difference between the free trade and constrained solutions is that the latter include flows from Alberta to eastern Canada while the former solutions do not. These flows from western to eastern Canada were available to be chosen in the free trade model. Their omission in the solution where final costs are minimized without border constraints suggests that their inclusion will raise the costs of meeting demand.

The difference in the value of the objective function between the constrained and free trade solutions ($z - Z$) lies within a very small range for all three price series used $50,439 to $50,636 per day.[5]

In the changes between a free trade solution and its corresponding constrained solution (see Tables 4.3 and 4.4), no field changes its total shipments. The total production costs must therefore be equal for each pair of solutions.[6] *The difference between the two solutions is caused by an*

---

4 / 'After the provincial regulatory board have approved removal of gas from the respective province, the National Energy Board considers the advisability of adequate protection of requirements for the total national use, prior to allowing the natural gas to flow south to the United States.' From a speech by J.W. Kerr, president, Trans Canada Pipelines Ltd., to the Institution of Gas Engineers (May, 1965).

5 / The volumes used (x_{ij}) are flow per day measures, the value of the objective function is a value per day.

6 / There is no theoretical reason why excess capacity between fields does not change when the constraints are changed. If excess capacity did shift between fields, the change in value of the objective function would include production costs also.

TABLE 4.4
Primal flows: constrained model, old contract prices, and combined prices, MMCFD

From \ To		BC 1	Alta 2	Sask. 3	Man. 4	Ont. NW 5	Ont. SW 6	Ont.–Tor. 7	Ont.SE 8	Quebec 9	US1 10	US2 11	US3 12	US4 13	US5 14	US6 15	US7 16	US8 17	US9 18	US10 19	Slack
British Columbia	1	219																		222	
Alberta	2		572	176		146	66	243	89	89								673		332	
Saskatchewan	3				103																
Ontario	4																				43
Mexico	5																				150
Louisiana North	6						121							1546							
Louisiana South	7											7960	2596	888			811				
Texas Gulf	8																10,662				
Texas Permian	9																2061		1406		83
Texas Midcontinent	10																1790				
Texas Panhandle	11															3250					
San Juan	12																		1397		
New Mexico SE	13																		1338		
Wyoming	14																	301	366		
Kansas	15													344	1978						
Oklahoma	16													3106		596					
Mississippi	17												429								
California	18																		1889		
West. Virginia	19										524	56									

TABLE 4.5
Primal flows: constrained model, new contract prices, MMCFD

From \ To		BC 1	Alta 2	Sask. 3	Man. 4	Ont. NW 5	Ont. SW 6	Ont.–Tor. 7	Ont. SE 8	Quebec 9	US1 10	US2 11	US3 12	US4 13	US5 14	US6 15	US7 16	US8 17	US9 18	US10 19	Slack
British Columbia	1	219																		222	
Alberta	2		572	176		146	66	243	89	89								678		382	
Saskatchewan	3				103																
Ontario	4																				43
Mexico	5																				150
Louisiana North	6						121							1546							
Louisiana South	7											7960	2595	888			728				83
Texas Gulf	8																10,662				
Texas Permian	9																2144		1406		
Texas Midcontinent	10																1790				
Texas Panhandle	11															3250					
San Juan	12																		1397		
New Mexico SE	13																		1338		
Wyoming	14																	301	366		
Kansas	15													344	1978						
Oklahoma	16													3105		596					
Mississippi	17												429								
California	18																		1889		
West virginia	19										524	56									

TABLE 4.6
Differences between constrained and free trade solutions, all price series, MMCFD

From \ To	5	6	7	8	9	13	16	17	18
2	146	66	243	89	89			(301)	(332)
6		(66)		(89)		155			
7						(301)	301		
9			(243)		(89)		(301)		633
14								301	(301)
15	(146)					146			

increase in transportation costs incurred in the constrained solution. The effect of constraining Alberta to ship to eastern Canadian markets creates an inefficient transportation network in North America with the end result that consumers in general pay more for natural gas.

When Alberta is forced to ship to eastern Canada, some other Albertan shipment must be correspondingly reduced. In all three solutions, the increased shipments from Alberta to eastern Canada are at the expense of shipments to California and US 8 (Montana, Wyoming, Utah, and Colorado). Californian demand is met by increased shipments from the Permian field of Texas while US 8 receives additional supplies from the Wyoming field.

An increase in flows, for example to California, must displace equivalent flows from these fields to other markets. The American producers who were supplying the eastern Canadian markets are now diverted to American markets. In all, six fields: Alberta, Louisiana North, Louisiana South, Permian Basin, Wyoming and Kansas, are affected. Nine demand points are affected: the five eastern Canadian markets and US districts 4, 7, 8, and 9.[7]

The increase in costs for the system as a whole due to the inclusion of these east-west flows can be disaggregated. The change in flows can be calculated at each of the demand points affected and these changes in flows can be valued at their respective costs. This is proved below.

The objective function for the free trade model is:

$$Z = a \sum_i \sum_j d_{ij} g_{ij} x_{ij} + \sum_i \sum_j p_i x_{ij}. \tag{2.8}$$

7 / These are basically the Chicago market, the Texas market, the Utah-Montana market, and the California market, respectively.

The objective function for this constrained solution is:

$$z = a \sum_i \sum_j d_{ij} g_{ij} \bar{x}_{ij} + \sum_i \sum_j p_i \bar{x}_{ij}, \tag{4.1}$$

where $z > Z$.

If we subtract 2.8 from 4.1 we get:

$$H = z - Z > 0, \tag{4.5}$$

$$H = a[\sum_i \sum_j d_{ij} g_{ij} \bar{x}_{ij} - \sum_i \sum_j d_{ij} g_{ij} x_{ij}] + \sum_i \sum_j p_i \bar{x}_{ij} - \sum_i \sum_j p_i x_{ij}. \tag{4.6}$$

Expanding H:

$$a[d_{11} g_{11} \bar{x}_{11} - d_{11} g_{11} x_{11} + \dots + d_{mn} g_{mn} \bar{x}_{mn} - d_{mn} g_{mn} x_{mn}] + [p_1 \bar{x}_{11} - p_1 x_{11} + \dots + p_m \bar{x}_{mn} - p_m x_{mn}]. \tag{4.7}$$

Without loss of generality, we can rearrange the elements in the first term of 4.7. Place all the elements first which involve a change in flow between the two solutions. There are y of these. The remaining $(mn - y)$ elements in the first term must be zero.

The second term in 4.7 can be rewritten as:

$$p_1 \left[\sum_{j=1}^{n} \bar{x}_{1j} + \sum_{j=1}^{n} x_{1j} \right] + p_2 \left[\sum_j \bar{x}_{2j} - \sum_j x_{2j} \right] + \dots + p_m \left[\sum_j \bar{x}_{mj} - \sum_j x_{mj} \right]. \tag{4.8}$$

Each bracketed term in 4.8 represents the sum of all the shipments of the jth field in the simulated solution. But each field supplies the identical amount in both solutions. Therefore the value of 4.8 is identically zero. 4.7 reduces to

$$H = a \left[\sum_{i=1}^{6} d_{i1} g_{i1} (\bar{x}_{i1} - x_{i1}) + \sum_{i=1}^{6} d_{i2} g_{i2} (\bar{x}_{i2} - x_{i2}) + \dots + \sum_{i=1}^{6} d_{i9} g_{i9} (\bar{x}_{i9} - x_{i9}) \right] \tag{4.9}$$

(remembering that terms have been rearranged so that fields and markets where flows change come first. Thus $i = 1, \dots, 6; j = 1, \dots, 9$ represents the six fields and nine markets where flows did change). Each term in 4.9 is the value of the change in flows at an affected demand point. The total increase in the value of the objective function $(z - Z)$ equals the sum of these increases in cost over all nine affected demand points.

Table 4.7 shows the changes in costs at each of the nine demand points.

TABLE 4.7
Changes in costs at each affected demand point due to inclusion of Canadian east-west flows

Demand point	Old prices		New prices		Combined prices	
	$/day	%	$/day	%	$/day	%
Ontario northwest	15,096	43.3	7023	16.3	15,096	43.3
Ontario southwest	8382	18.1	6673	15.2	7372	15.9
Ontario-Toronto	16,548	21.9	14,118	18.1	15,844	20.8
Ontario southeast	8651	33.4	7769	21.8	7289	26.7
Quebec	5438	18.1	4548	14.7	5180	17.1
Subtotal Canada	54115	25.4	40131	17.1	50780	23.3
USA4	(7530)	0.5	(11,363)	0.3	(12,333)	0.9
USA7	(8699)	—	(2288)	—	(2370)	—
USA8	(12,612)	6.0	(5569)	3.0	(9451)	4.5
USA9	25,335	1.8	29,514	1.8	24,010	1.6
Subtotal USA	(3506)	—	10,294	—	(144)	—
Total	50,609		50,425		50,636	

All five markets in eastern Canada exhibit large increases in costs when Alberta is forced to ship through to these markets. Consumers in eastern Canadian markets pay between $40,130 to $54,115 more per day (depending on the estimates of the production costs in the USA) when imports are restricted. These cost increases range from 15 per cent in the Quebec market to over 43 per cent in Ontario Northwest. On average the eastern Canadian consumer pays between 17 and 25 per cent more to get gas from Alberta rather than from a source south of the border.

For two of the three columns in Table 4.7 (2 of the 3 alternative estimates of USA production costs), American consumers on the whole are *better off* when Alberta is constrained to meet demand in eastern Canadian markets.

The Canadian policy of Canadian gas for Canadian consumers, a policy which substantially increases costs to eastern Canadian consumers, may have as one of its indirect effects, the reduction in costs for American consumers.

In all three cases, American markets 4 (Chicago) and 8 (Utah-Montana) are better off when American imports are prevented from penetrating Canadian markets. These rather strange results can be explained.[8]

8 / This does not contradict the free trade theories in international trade where both countries are better off after trade. That theory deals with two extreme cases: no trade and perfectly free trade. The solutions in this paper begin with some trade. It is not obvious that moving from some trade to free trade will make both parties better off.

Final gas costs to American consumers are *lower*, the *more* self sufficient Canada becomes. Canadian markets are far apart, strung on a thin line from the Pacific to the Atlantic. They are therefore expensive markets to serve. In a North American free trade gas market, these Canadian markets would be served at lowest cost commensurate with lowest total North American costs. In the case of the free trade solutions, eastern Canada is supplied by Kansas, North Louisiana, and the Permian Basin. Markets close to Canada such as Chicago and Wisconsin in the North American free market case are supplied by Southern Louisiana which is a more expensive supplier in total than North Louisiana or Kansas. When Alberta is constrained to ship to eastern Canadian markets, Wisconsin and Chicago can be served by Northern Louisiana and Kansas.

A policy of Canadian gas for Canadian demand means that American supplies which would go to Ontario in a free trade model can replace more expensive shipments which were supplying the American mid-west in the free trade model.[9]

In Table 4.8, the tableau of payments is presented for the free trade and constrained solution as is their difference. The price set used is that combining the old and new contract prices.

The first row shows, for example, that Canadian consumers in the free trade solution pay $137,631 to Canadian producers, $105,478 to American producers, and $171,659 in marginal transportation costs per day for all the gas they consume. Similarly in the constrained solution, American consumers pay $161,462 to Canadian producers, $6,881,376 to American producers, and $3,747,285 in transportation costs per day.

Subtracting the free trade from the constrained trade payments gives the changes in payment patterns. The total value of the objective function increases by $50,636. The total increase of $50,780 to Canadian consumers is made up of $42,036 in increased transportation charges and $8744 in payments to producers. The net gain to Americans of $144

9 / If one could believe that the FPC was either rational or used linear programming, its seemingly erratic behaviour could be explained – refusing export and import licences but quickly approving a line which would serve Canadian markets from Canadian fields via a pipeline through the USA (Great Lakes Transmission Company). All these acts of course increase Canadian self-sufficiency; increases in Canadian demand are met by Canadian production. The FPC, knowing the facts of northern life and maximizing the welfare of American consumers, undertook to maximize Canadian self-sufficiency.

I do not have great faith in the explanatory power of this footnote.

TABLE 4.8
Final payments for natural gas, $/day

	Producers		Transportation costs	Total
	Canadian	American		
Free trade solution				
Canadian consumers	137,631	105,478	171,659	414,768
American consumers	243,622	6,807,960	3,738,685	10,790,267
Total	381,253	6,913,438	3,910,344	11,205,035
Constrained solution				
Canadian consumers	219,791	32,062	213,695	465,548
American consumers	161,462	6,881,376	3,747,285	10,790,123
Total	381,253	6,913,438	3,960,980	11,255,671
Constrained–free trade				
Canadian consumers	82,160	(73,416)	42,036	50,780
American consumers	(82,160)	73,416	8600	(144)
Total			50,636	50,636

is made up of an increase in transportation charges of $8600 and a decrease in payments to producers of $8744.

Table 4.7 gave the *net* increases in costs at demand points when Alberta was constrained to ship into eastern Canadian markets. These increased costs represented changes in both payments to producers and transportation charges.

4 SUMMARY

In the beginning of this chapter, the solutions to the free trade model were presented. Comparing the flows in these solutions with the actual flows showed significant deviations in movements across the border. A second model (the constrained) was then developed to simulate the actual flow pattern.

Comparing the free trade and constrained solutions yielded estimates of the increase in costs for North America as a whole due to the east-west transcontinental gas flows. Disaggregating this cost for the system to individual markets showed that eastern Canadian consumers bore 80 to 107 per cent of these costs. A move in 1966 to free trade in natural gas would greatly benefit eastern Canadian consumers but might hurt American gas users.

The cost of the actual inefficient transportation pattern amounts to

$18,400,000 per year. The present value of the future stream of such costs discounted at 10 per cent is $184,000,000.

The decision to restrict trade in natural gas between the two countries has cost consumers in North America an additional $184,000,000. Eastern Canadian consumers are forced to pay between $146,000,000 (new contract prices in the USA) and $197,000,000 (old contract prices) more for natural gas than they would under free trade.[10] American consumers have either been hurt by paying $34,000,000 more over time (new prices), or have gained $12,800,000 (old contract prices).

Authorities must then consider whether the benefits of restricted trade are in the neighbourhood of the costs – $180,000,000. Moreover, if national defence, self-sufficiency, or unconscious pride is worth this amount, are trade restrictions of natural gas, the best or the cheapest means of obtaining these benefits?

---

10 / I have ignored the increase in capital costs incurred by Canadians because of the decision to build through the muskeg north of Lake Superior. As n. 21 in chapter 1 showed, a lower bound estimate of the increased costs incurred in building north instead of south of the Great Lakes was $35,000 per mile or $22,000,000. Total increased costs to Canada then ranged between $166,000,000 and $217,000,000 to have east-west gas flows which travel north of the border. No attempt has been made to estimate the difference in capital costs between supplying eastern Canada from Alberta or the USA.

5

The dual

1 MODEL

To each primal model described in chapters 2 and 4 there corresponds a dual model. For each equation in the constraint set of the primal, a new variable is introduced. For each of the first n constraints (which are the demand inequalities), a variable u_j ($j = 1, \dots, n$) is introduced. Variable v_i ($i = n + 1, \dots, n + m$) is introduced for each of the last m constraints (the supply inequalities).

The dual is in the form:

$$\text{maximize } w = \sum_{j=1}^{n} u_j D_j - \sum_{i=1}^{m} v_i S_i, \tag{5.1}$$

$$\text{subject to } u_j - v_i \leq c_{ij} \quad i = 1, \dots, m, j = 1, \dots, n, \text{ where} \tag{5.2}$$

$$c_{ij} = [ad_{ij}g_{ij} + p_i].$$

The right hand side of 5.2 is the total average costs (in dollar units) of meeting a unit of demand j from field i. The left hand side of 5.2 must also be in dollar units. u_j (the shadow price at the market) can be interpreted as the delivered price of the marginal unit of gas in the jth market. v_i (the shadow price at the field) can be considered to be the royalty rent earned by the marginal unit of field production in field i.[1]

---

1 / c_{ij}, the average costs of activity x_{ij}, include average production costs. These average production costs include an average rate of return. v_i is thus the return to the field above the average return, i.e., the royalty rent of the field.

u_jD_j represents the total costs at the market j of meeting that market's demand. From the producers' standpoint, u_jD_j represents the total gross revenue received in market j. The sum of u_jD_j over all j ($j = 1, ..., n$) is then the total gross revenue received by all the suppliers.

v_iS_i is the product of the total output of a field i and the royalty rental value of the marginal unit of i's production. v_iS_i is the royalty rent value of the production at area i. The sum of v_iS_i over all i ($i = 1, ..., m$) is the total royalty value of the total system field production.

The objective function of this dual solution is to maximize w, the difference between the gross revenue of the fields ($\sum_j u_jD_j$) and the rental received by these fields ($\sum_i v_iS_i$). w can be considered as the net revenue of the fields, i.e., revenue net of rental payments.

The second constraint states that for any supply point i and any demand point j, the unit delivered price in the market j (u_j) is to be no greater than the sum of the actual costs of delivery from the field i (c_{ij}) plus the unit royalty earned by field i (v_i).

The dual maximizes the net revenue of all the fields subject to the constraint that no supplier earns a profit above the sum of his average costs of delivery and the unit royalty value of his field.

2 DUALITY THEOREM

The duality theorem[2] establishes three major points.

(i) Where the primal has an optimal feasible solution Z, the dual has an optimal feasible solution w, and these two are identical.[3]

(ii) Where a flow x_{ij} occurs at a positive level in the primal solution, that delivery earns zero positive profits in the dual solution, i.e.,

$$u_j - v_i = c_{ij}; \text{ if } x_{ij} > 0; \tag{5.3}$$

where flow x_{ij} is at zero level in the primal solution, that delivery earns negative profits in the dual solution,

$$u_j - v_i < c_{ij}\ ; \text{ if } x_{ij} = 0. \tag{5.4}$$

2 / One representation is given in Dantzig (1963), pp 120–144.

3 / In a competitive system, minimizing costs or maximizing revenue will lead to the same solution.

(iii) If S_i, the capacity of field i, is fully utilized in the primal solution, field i earns a positive royalty in the dual solution. If the capacity of field i is not completely used, that field earns no royalty rent.

$$v_i > 0 \text{ if } -\sum_{j=1}^{n} x_{ij} = -S_i, \tag{5.5}$$

$$v_i = 0 \text{ if } -\sum_{j=1}^{n} x_{ij} \geq -S_i. \tag{5.6}$$

Assume for the moment that the total demand of the system's markets just equals that system's field capacity. *If there is no excess capacity in the system, the dual solution is indeterminate.*

Intuitively this can be proven as follows. Since total supply is just sufficient to meet demand, each field must supply its entire output *whatever* prices may be. Raising a field's activity costs will not remove it from the basic solution. Changing relative costs will change the flow pattern but it cannot decrease the supply which a field ships. A more rigorous proof can be constructed. The dual objective function maximizes the difference between two sets of unknowns (n demand variables, and m supply variables).

$$\text{maximize } w = \sum_{j=1}^{n} u_j D_j - \sum_{i=1}^{n} v_i S_i \tag{5.7}$$

subject to a set of mn constraints

$$u_j - v_i \leq c_{ij} \quad (\text{all } i, i = 1, \ldots, m; \text{ all } j, j = 1, \ldots, n). \tag{5.8}$$

The duality theorem states that the number of positive flows in the primal will equal the number of *independent* constraints in the primal. A constraint in 5.8 will be satisfied by an equality, if and only if the flow associated with cost c_{ij} is at a positive level in the primal (from 5.3). Therefore, the number of constraints satisfied by an equality in the dual equals the number of *independent* constraints in the primal.

Where the total capacity of the system is just equal to the total system demand, it can be shown that if there are $n + m$ constraints in the primal formulation, only $n + m - 1$ of these are independent. Suppose that the first n of the primal constraints are satisfied. Demand is met in every market. If then $m - 1$ of the supply constraints are met, the mth constraint must also be met. Specifically the mth constraint is a linear combination of the other $n + m - 1$ constraints.[4] What the mth field supplies must be the sum of all the market demands less the sum of the supplies from the other $m - 1$ fields.

The dual problem in this case where the total available capacity equals the total system demand consists of finding values for $n + m$ variables when there are only $n + m - 1$ independent equations.[5] There is no unique solution to this problem.

While the absolute levels of the dual variables are indeterminate, their relative values are determinate.

For two fields i and h which both deliver to market j, the dual constraints are:

$$u_j = c_{ij} + v_i \tag{5.9}$$

$$u_j = c_{hj} + v_h, \text{ therefore} \tag{5.10}$$

$$c_{ij} + v_i = c_{hj} + v_h \tag{5.11}$$

$$v_i - v_h = c_{hj} - c_{ij}. \tag{5.12}$$

The difference between the royalties of two fields supplying the same market must equal the difference in the costs of supplying that market from the two fields. This difference is determinate, since the costs of supplying a market from any field must be determinate. However, there are an infinite number of combinations of v_i and v_h that will yield this difference.[6]

Where excess capacity does exist for at least one field (i.e., total field capacity is greater than total market demand), that excess capacity determines unique dual values for the system.

---

4 / Let $D_1{}^c, ..., D_n{}^c$ stand for the demand constraints, $S_1{}^c ..., S_m{}^c$ stand for the supply constraints; then $S^c{}_m = 1.D_1{}^c + 1.D_2{}^c + ... + 1.D_n{}^c - 1.S^c{}_{n+1} - 1.S^c{}_{n+2} - ... - 1.S^c{}_{n+m-1}$.

5 / The equating of the number of unknowns to the number of independent equations is neither a necessary nor a sufficient condition for the uniqueness of a solution. Generally, where the number of independent equations is less than the number of unknowns, it is not possible to find a unique solution to the problem.

6 / What are $v_a, v_b, v_c, v_d, ...$? They are really nothing but determinate arbitrary terms that have meaning only in their proportionate relationship to one another ... since there can be no more than m raretés for each trader in a market of m commodities, there are at most only m indeterminate r terms expressing the values in exchange of the m commodities, when the market is in a state of general equilibrium. These terms, taken two at a time, yield $m\ (m - 1)$ prices of the m commodities in terms of one another. Walras (1965), 180.

Three sets of prices were used in finding free trade solutions. In this example, for purposes of illustration only, the dual values related to the first or third price sets, i.e., the old contract price series or the set which is a combination of old and new prices, will be used. The conclusions drawn can of course be obtained in a similar manner from the second price series, the one for contracts in the USA signed after 1961.

In the free trade solution, southern Louisiana and Ontario have excess capacity. Southern Louisiana ships 95 per cent of its capacity. Ontario ships none of its supply.

Consider a market j, where Louisiana South (1) ships:

$$x_{1j} > 0, \tag{5.13}$$

therefore, from 5.4,

$$u_j - v_1 = c_{1j}. \tag{5.14}$$

But the total shipments from field 1 are less than the total capacity at 1;

$$\sum_{j=1}^{n} x_{1j} \leq S_1 \tag{5.15}$$

therefore, from 5.6

$$v_1 = 0. \tag{5.16}$$

Southern Louisiana earns *no* royalty.

Therefore, from 5.14

$$u_j = c_{1j} \tag{5.17}$$

The delivered prices in the markets to which southern Louisiana ships gas must be equal to the costs of supplying that market from southern Louisiana.

If another field k ships gas to the same market as southern Louisiana and if k has no excess capacity, the costs of supplying market j must be less for k than for 1 (southern Louisiana). Where southern Louisiana supplies a market in conjunction with other fields,[7] it must be the marginal supplier to that market. The delivered price to the market j is uniquely determined by southern Louisiana.[8]

---

7 / If southern Louisiana were the *sole* supplier in the markets it served, this development would break down. The system in this case would be separable into two non-interacting systems, one supplied solely by southern Louisiana, the other being supplied by the rest of the fields.

8 / The existence of excess capacity for the system destroys the dependence of the $(n+m)$th equation in the primal on the other $m+n-1$. The introduction of excess capacity yields a system of $n+m$ equations in $n+m$ unknowns.

$$u_j = c_{1j}, \text{ also} \tag{5.17}$$

$$u_j = c_{kj} + v_k. \tag{5.18}$$

The delivered price in market j must be equal to the costs of delivery from field k plus the positive royalty earned by k.

$$(5.18) - (5.17): v_k = c_{1j} - c_{kj}. \tag{5.19}$$

The supplier k must earn a positive royalty (v_k) equal to his cost advantage over southern Louisiana ($c_{1j} - c_{kj}$).

Assume that Louisiana south is not the most expensive supplier to market j. Southern Louisiana has excess capacity, therefore the shipment 1 to j can be increased by one unit without correspondingly decreasing any other shipment 1 to q ($q \neq j$). Since Louisiana south earns no positive royalty, the increased shipment would cost the market c_{1j}. Demand at j would now be oversupplied by one unit. Since costs are positive, the value of the primal objective function is lowered (a good move) if this one extra unit is not shipped. Let field k have the same costs of delivery to market j as field 1. Decrease the shipments from k to j by one unit. All demands are being met in each market. Assume that k is the most expensive supplier to all other markets.[9] The shipments to all other fields are not affected. No market outside j is affected. Field k now has excess capacity of one unit. It therefore earns a zero royalty where before it earned a positive royalty v_k.

If southern Louisiana replaces k in supplying one unit of demand at market j, that market would save v_k. Therefore k could not have had the same costs of supplying the market as 1. Southern Louisiana must be the most expensive supplier.

Consider another market, r, where field k ships, but southern Louisiana does not. Another field m ships to market r.

We know

$$u_r = c_{kr} + v_k, \text{ and} \tag{5.20}$$

$$u_r = c_{mr} + v_m. \tag{5.21}$$

But from 5.19

$$v_k = c_{1j} - c_{kj}. \tag{5.19}$$

---

9 / This assumption is merely a simplifying one. The proof can easily be extended to the case where k replaces some other supplier.

Substituting 5.19 in 5.20 we get

$$u_r = c_{kr} + c_{1j} - c_{kj}. \tag{5.22}$$

Subtracting 5.22 from 5.21 and solving for v_m yields:

$$v_m = (c_{kr} - c_{mr}) + (c_{1j} - c_{kj}). \tag{5.23}$$

The royalty earned by the *m*th field (v_m) equals the cost advantage of the *m*th field over the *k*th field in the *r*th market plus the cost advantage of the *k*th field over Louisiana south in the *j*th market.[10]

The initial portion of this chapter showed that where no excess capacity exists in the system, the field royalties can only be determined in relation to one another.

$$v_i - v_h = c_{hj} - c_{ij}. \tag{5.12}$$

The absolute values of the dual variables are indeterminate. When excess capacity exists in a system, however, some v_1 is zero. In this case v_i is determinate and therefore all other v_h are determinate also.

3 THE CHANGES IN THE DUAL VARIABLES FROM THE FREE TRADE TO THE CONSTRAINED SOLUTION

Tables 5.1 and 5.2 give the dual variables at fields (v_i) and markets (u_j) for both the free trade and constrained solutions.[11] Of the nineteen fields, three in Canada have an increase in the shadow price of \$0.0144, when

10 / This analysis can be composed in profit maximizing terms. The above analysis has field *k* shipping to markets *r* and *j*, while field *m* ships only to *r*. Therefore,

$x_{mj} = 0$, therefore, from 5.4
$u_j - v_m < c_{mj}$, but
$v_m = c_{kr} - c_{mr} + v_k$, therefore
$u_j - (c_{kr} - c_{mr}) - v_k < C_{mj}$, but
$u_j = c_{kj} + v_k$, therefore
$c_{kj} - c_{mj} < c_{kr} - c_{mr}$.

Field *m* does not ship to *j* because its cost advantage in the *j*th market is less than its cost advantage in the *r*th market. Profit maximization ensures that *m* will serve *r* and not *j*.

11 / Remember that these dual values are dependent on the primal prices used.

TABLE 5.1
Dual values at the fields, $/MCFD

Field		Old price series			Combined prices			New price series		
		Free trade	Constrained	Difference	Free trade	Constrained	Difference	Free trade	Constrained	Difference
1	British Columbia	0.0078	0.0261	0.0143	0.0146	0.0290	0.0144	0.0394	0.0538	0.0144
2	Alberta	0.0133	0.0277	0.0144	0.0162	0.0306	0.0144	0.0410	0.0553	0.0143
3	Saskatchewan	0.0250	0.0394	0.0144	0.0280	0.0423	0.0143	0.0527	0.0671	0.0144
4	Ontario	—	—	—	—	—	—	—	—	—
5	Mexico	—	—	—	—	—	—	—	—	—
6	Louisiana N	0.0400	0.0400	—	0.0280	0.0280	—	0.0417	0.0417	—
7	Louisiana S	0.0289	0.0289	—	0.0080	0.0080	—	—	—	—
8	Texas Gulf	0.0055	0.0055	—	0.0038	0.0038	—	0.0517	0.0517	—
9	Permian	—	—	—	—	—	—	0.0076	0.0076	—
10	Midcontinent	0.0393	0.0394	—	0.0402	0.0402	—	0.0587	0.0587	—
11	Texas Panhandle	0.0443	0.0443	—	0.0270	0.0270	—	0.0255	0.0255	—
12	San Juan	0.0635	0.0635	—	0.0664	0.0664	—	0.0911	0.0911	—
13	New Mexico SE	0.0149	0.0149	—	0.0128	0.0128	—	0.0224	0.0224	—
14	Wyoming	0.0695	0.0695	—	0.0619	0.0619	—	0.0738	0.0738	—
15	Kansas	0.0688	0.0688	—	0.0717	0.0717	—	0.0382	0.0382	—
16	Oklahoma	0.0819	0.0819	—	0.0538	0.0538	—	0.0493	0.0493	—
17	Mississippi	0.0520	0.0520	—	0.0336	0.0336	—	0.0134	0.0134	—
18	California	0.0377	0.0377	—	0.0406	0.0406	—	0.0654	0.0654	—
19	West Virginia	0.0322	0.0322	—	0.0192	0.0192	—	0.0311	0.0311	—

TABLE 5.2
Dual values at the markets, $/MCFD

	Market	Old price series			Combined prices			New price series		
		Free trade	Constrained	Difference	Free trade	Constrained	Difference	Free trade	Constrained	Difference
1	British Columbia	0.2172	0.2316	0.0144	0.2202	0.2346	0.0144	0.2449	0.2593	0.0144
2	Alberta	0.1851	0.1995	0.0144	0.1881	0.2025	0.0144	0.2128	0.2272	0.0144
3	Saskatchewan	0.1949	0.2093	0.0144	0.1979	0.2123	0.0144	0.2226	0.2370	0.0144
4	Manitoba	0.2359	0.2503	0.0144	0.2388	0.2532	0.0144	0.2636	0.2780	0.0144
5	Ontario NW	0.3079	0.3700	0.0621	0.3107	0.3729	0.0622	0.3355	0.3977	0.0622
6	Ontario SW	0.2875	0.4022	0.1147	0.2904	0.4051	0.1147	0.3151	0.4299	0.1148
7	Ontario–Toronto	0.3103	0.4062	0.0959	0.3133	0.4091	0.0958	0.3380	0.4339	0.0959
8	Ontario SE	0.3312	0.4162	0.0850	0.3342	0.4191	0.0849	0.3589	0.4439	0.0850
9	Quebec–Montreal	0.3385	0.4271	0.0886	0.3412	0.4299	0.0887	0.3660	0.4547	0.0887
10	US1	0.3623	0.3623	—	0.3652	0.3652	—	0.3899	0.3899	—
11	US2	0.3285	0.3285	—	0.3314	0.3314	—	0.3561	0.3561	—
12	US3	0.2382	0.2382	—	0.2411	0.2411	—	0.2658	0.2658	—
13	US4	0.2535	0.2535	—	0.2564	0.2564	—	0.2811	0.2811	—
14	US5	0.2255	0.2255	—	0.2283	0.2283	—	0.2531	0.2531	—
15	US6	0.2023	0.2023	—	0.2052	0.2052	—	0.2300	0.2300	—
16	US7	0.1925	0.1925	—	0.1954	0.1954	—	0.2262	0.2262	—
17	US8	0.2244	0.2389	0.0144	0.2275	0.2418	0.0144	0.2522	0.2666	0.0144
18	US9	0.3187	0.3187	—	0.3216	0.3216	—	0.3464	0.3464	—
19	US10	0.2689	0.2833	0.0144	0.2719	0.2863	0.0144	0.3110	0.3254	0.0144

Alberta is forced to ship to eastern Canadian markets. The marginal delivered prices at eleven of the nineteen markets increase from the free trade to the constrained solutions. Of these eleven changes, six are identical to each other and equal to the change in royalty value of the fields.

An explanation of this odd result is in order. First, note that the fields and markets where shadow prices change are not the same as those which experience a change in their flows between the two solutions.

These increases in the value of the dual variables can be ordered in three ways. After noting these three points, and with the aid of the development at the beginning of the chapter, the unusual result can be easily explained.

(i) Only the three western Canadian fields receive higher royalties in the constrained solution. The only one of these three to also experience a change in flows is Alberta.

(ii) On page 55, it was noted that in the primal solution to the free trade model, there appears to be a natural dividing line between eastern and western North America.[12] The delivered cost for each of the seven western markets except California, increases by \$0.0144. The only change in flows, however, occur in the Wisconsin market (17) and California (18).

(iii) All the other markets which have an increase in their delivered costs are the eastern Canadian markets where Alberta is constrained to ship. These markets' shadow prices increase by more than \$0.0144.

The key would therefore appear to be in the only western field and market which experience both a change in flow and a change in dual value – field 2 (Alberta) and market 17 (see Table 5.3).

In the constrained solution, Alberta is forced to ship to eastern Canadian markets. The increase in Alberta's Canadian shipments is partially offset by a reduction of its shipments to market 17 in the USA from 970 to 679 MMCFD. This reduction in supply to market 17 is compensated by a shipment from field 14 to market 17.

From page 70, a constraint can be written as

$$u_j = c_{ij} + v_i. \tag{5.9}$$

12 / The western section consists of British Columbia, Alberta, Saskatchewan, Manitoba, and US districts 8, 9, and 10.

TABLE 5.3
Changes in flows and in dual values between the free trade and constrained solutions (old and combined price series)

Field	Change in flow (yes, no)	Change in dual value ($/MCFD)
1 British Columbia	No	0.0144
2 Alberta	Yes	0.0144
3 Saskatchewan	No	0.0144
6 Louisiana North	Yes	—
7 Louisiana South	Yes	—
9 Texas-Permian	Yes	—
14 Wyoming	Yes	—
15 Kansas	Yes	—
Market		
1 British Columbia	No	0.0144
2 Alberta	No	0.0144
3 Saskatchewan	No	0.0144
4 Manitoba	No	0.0144
5 Ontario NW	Yes	0.0622
6 Ontario SW	Yes	0.1147
7 Ontario–Toronto	Yes	0.0758
8 Ontario SE	Yes	0.0849
9 Quebec–Montreal	Yes	0.0987
14 US5	Yes	—
15 US6	Yes	—
17 US8	Yes	0.0144
19 US10	No	0.0144

The constraints for market 17 and fields 2 and 14 can be written as:

Free trade solution

$$x_{2,17} > 0, \therefore u_{17} = c_{2,17} + v_2, \tag{5.24}$$

$$x_{14,17} = 0, \therefore u_{17} < c_{14,17} + v_{14}. \tag{5.25}$$

Constrained solution

$$\bar{x}_{2,17} > 0, \therefore \bar{u}_{17} = c_{2,17} + \bar{v}_2 \tag{5.26}$$

$$\bar{x}_{14,17} > 0, \therefore \bar{u}_{17} = c_{14,17} + \bar{v}_{14} \tag{5.27}$$

Therefore (5.27 less 5.24):

$$\bar{u}_{17} - u_{17} = c_{14,17} + \bar{v}_{14} - c_{2,17} - v_2, \tag{5.28}$$

but $\bar{v}_{14} = v_{14}$ (Table 5.2), therefore

$$\bar{u}_{17} - u_{17} = c_{14,17} - c_{2,17} + v_{14} - v_2. \tag{5.29}$$

When Alberta is replaced by Wyoming as a supplier to market 17, the dual variable for that market must rise since Wyoming is a more expensive supplier than Alberta. At the shadow price which ruled in the free trade solution, Wyoming was too costly for market 17 ($u_{17} < c_{14,17} + v_{14}$).

To induce Wyoming to supply that market, price must rise ($\bar{u}_{17} = c_{14,17} + v_{14}$). At this new price ($\bar{u}_{17}$), Alberta also supplies the market. Therefore Alberta must be offered a higher price and the change in royalty value for Alberta must equal the increased costs in having market 17 supplied by Wyoming instead of Alberta.

$$\bar{v}_2 - v_2 = \bar{u}_{17} - u_{17} = c_{14,17} - c_{2,17} + v_{14} - v_2. \tag{5.30}$$

In market 18, a similar change in flows occurs – Alberta is replaced by an American source. Unlike the changes in market 17, the shadow delivered price in California does not change ($u_{18} = u_{18}$).

The explanation of this result is as follows:

Free trade solution

$$x_{2,18} > 0, \therefore u_{18} = c_{2,18} + v_2, \tag{5.31}$$
$$x_{9,18} > 0, \therefore u_{18} = c_{9,18} + v_9, \tag{5.32}$$
$$x_{14,18} > 0, \therefore u_{18} = c_{14,18} + v_{14}. \tag{5.33}$$

Constrained solution

$$x_{2,18} = 0, \therefore \bar{u}_{18} < c_{2,18} + \bar{v}_2, \tag{5.34}$$
$$x_{9,18} > 0, \therefore \bar{u}_{18} = c_{9,18} + \bar{v}_9, \tag{5.35}$$
$$x_{14,18} > 0, \therefore \bar{u}_{18} = c_{14,18} + \bar{v}_{14}. \tag{5.36}$$

In the free trade solution, fields 2, 9, and 14 supply California, while in the constrained solution, field 2 no longer supplies the market and field 9 increases its shipments.

But $\bar{v}_9 = v_9 = 0$ (field 9 has excess capacity) and $\bar{v}_{14} = v_{14}$. Therefore,

$$\bar{u}_{18} = u_{18} = c_{9,18}. \tag{5.37}$$

The shadow delivered price in market 18 equals the cost of supplying the market from field 9.

From (5.36 – 5.35)

$$\bar{v}_{14} = v_{14} = c_{9,18} - c_{14,18} \tag{5.38}$$

The royalty rent of field 14 is determined by the difference in costs between fields 14 and 9 in supplying California.

$$\bar{v}_2 > v_2 = c_{9,18} - c_{2,18}.\text{[13]} \tag{5.39}$$

13 / The difference between shadow prices for market 17 can now be simplified.

$u_{17} = c_{2,17} + v_2 = c_{2,17} + c_{9,18} - c_{2,18}$.

$\bar{u}_{17} = c_{14,17} + v_{14} = c_{14,17} + c_{9,18} - c_{14,18}$, therefore

$\bar{u}_{17} - u_{17} = c_{14,17} - c_{2,17} + c_{2,18} - c_{14,18}$. Similarly

$$\bar{v}_2 - v_2 = \bar{u}_{17} - u_{17} = c_{14,17} - c_{2,17} + c_{2,18} - c_{14,18}. \tag{5.a}$$

The royalty value of Albertan production in the free trade model was equal to the difference in the costs of supplying California from Alberta and field 9. However, the increase in the royalty value of Alberta because of the changes in flows to market 17, shuts Alberta out of the California market as it was too costly a supplier in the constrained solution.

Since the royalty value of Alberta increases in the constrained solution, all the markets which it supplies in both solutions must also experience an increase in their shadow price. Therefore, dual prices increase by $0.0144 at markets 2, 3, 19. Since field 1 (British Columbia) supplies market 19 (Pacific northwest) in both solutions, and the shadow delivered price at market 19 increases by $0.0144, so the royalty value in British Columbia fields rises by $0.0144. Likewise, the shadow delivered price in British Columbia markets must increase by this same amount.

The change in prices for field 3 and market 4 is an anomaly since in both solutions field 3 is the only supplier to market 4 and ships all its production there. Technically, the shipment from field 3 to market 3 was not zero (see p. 52, n. 1) but positive. Fields 2 and 3 are such good substitutes in market 3, that the royalty value of field 3 rises in response to increases in dual values at market 3.

Delivered prices in all eastern Canadian markets rise by an amount greater than $0.0144. These delivered prices must rise by an amount just sufficient to induce Alberta suppliers to ship there.

In the free trade solution, Alberta does not ship to eastern Canada:

$$x_{2,j} = 0 \quad (j = 5, ..., 9), \text{ therefore} \tag{5.40}$$

$$u_j < c_{2,j} + v_2 \quad (j = 5, ..., 9). \tag{5.41}$$

The eastern Canadian delivered price in the free trade solution is not sufficient to meet the costs of shipping from Alberta.

In the constrained solution:

$$\bar{x}_{2,j} > 0 \quad (j = 5, ..., 9) \tag{5.42}$$

$$\bar{u}_j = c_{2,j} + \bar{v}_2 \quad (j = 5, ..., 9). \tag{5.43}$$

The increase in the delivered price in eastern Canada was due to *two* influences. If the royalty value at Alberta did *not* increase, the shadow price at market j would have to rise by:

$$\bar{u}_j - u_j = (c_{2,j} + v_2) - u_j. \tag{5.44}$$

However, the royalty value at Alberta does increase:[14]

$$\bar{v}_2 - v_2 = c_{14,17} - c_{2,17} + c_{2,18} - c_{14,18}. \qquad (5.a)$$

Therefore, the eastern Canadian delivered price must rise by:

$$\bar{u}_j - u_j = (c_{2,j} + v_2) - u_j + c_{14,17} - c_{2,17} + c_{2,18} - c_{14,18} \quad (j = 5, \ldots, 9). \qquad (5.45)$$

These increases in delivered prices are necessary to induce a flow at a positive level in the constrained solution, where that flow was at a zero level in the free trade solution. The flow was zero in the latter solution because it would have operated at a loss. There was a cheaper source available to meet eastern Canadian demand. The increase in delivered prices (shadow prices) in eastern Canadian markets between the free trade and constrained solutions was necessary in order to have Alberta serve these markets without incurring a loss. These increases in shadow prices at these markets must then represent the cost advantage of American suppliers over Alberta under free trade conditions in eastern Canadian markets.

4 BENEFITS TO PRODUCERS

In chapter 1, I discussed the rise of regulation in the oil and gas industry. While some regulations can be considered to be in the consuming public's interest, it is not unreasonable to suggest that other regulations can improve the lot of producers. In the developments of the industry, the play and counterplay between competing interest groups has resulted in a set of regulations which do not necessarily maximize the welfare of one group. Instead, the compromises trade off gains of one group against

---

14 / The replacement of American suppliers by Alberta in eastern Canadian markets cannot *in this formulation increase* the rent earned by Alberta. The cost minimization procedure ensures that costs at j will increase only by the minimum necessary to induce Alberta to ship to j. (Note that where the Permian Basin replaces Alberta, the Permian field does not earn an increased rent.)

In the real world, guaranteeing Alberta the eastern Canadian market can easily lead to monopoly profits for the Alberta fields. These monopoly profits are limited when there is interfuel competition. The government policy for other fuels is similar to their gas policy, eastern Canadian *fuel* markets being reserved to a large extent for western Canadian producers. It is likely that monopoly profits are being earned by western Canadian producers.

losses. In total, however, some groups can be better or worse off, perhaps out of shear fortune rather than through some contrived master plan.

In chapter 4, the losses to consumers were detailed. In total, taking 1966 as an average year, the consuming public would have been better off were producers allowed to engage in free trade.[15]

Regulations which set the maximum prices to be charged cannot benefit producers. Similarly, preventing producers from selling to the highest bidder, appears at first glance to involve a cost rather than a benefit to producers. The prevention of gas exports may benefit producers, however, if imports are also prohibited, yielding a monopoly in the home market.

The linear programming model can be used to indicate the benefits to producers of trade restrictions. However, the static assumptions of the model make this measure of benefits rather weak. Producers will have benefitted from an exportable surplus policy if the gains from selling in a protected market compensate for the losses in selling in foreign markets. The losses in foreign markets consists of the price which would have been set and the volume sold. Under free trade, price, volume sold, and expectations would determine the exploration rate. To use actual production under restrictions as the estimate of the possible foreign sales under free trade is naive. In this section the benefits to producers are then overstated. No allowance is made for the losses to producers of volumes of gas which they did not produce in 1966 but which might have been forthcoming under free trade.

From Table 5.1, it can be observed that all producers are equally well off under constrained trade as they are under free trade except for the three western Canadian producers who are better off. The royalty value of production in western Canada increases by $0.0144 per MCF when imports into Canada are restricted.

While these shadow royalty values are not a direct monetary flow, they indicate the value to the producer of an extra unit of production. Constraining Alberta to meet eastern Canadian demand increases the value of production in western Canada by $34,018 per day[16] or $12,417,000 on an annual basis.

---

15 / The consuming public consists of a large number of conflicting groups itself. No attempt is made, for example, to estimate the costs or benefits of industrial versus commercial users.

16 / This value is found by multiplying capacity in western Canada by the change in the shadow value of field production.

The decision to restrict imports into Canada increases the present value of field capacity in western Canada by $124,000,000 (discounting at 10 per cent).

Stigler's maxim noted in the first chapter (that regulation of industries invariably benefits those industries) holds for natural gas.

To the hypotheses which have have been given traditionally to explain the presence of the Canadian transcontinental gas pipeline – national defence, etc. – can be added another, profits of producers.

6
The tariff

1 SUMMARY

From 1953 through 1968,[1] a tariff of 3¢ per MCF was levied on all imports of natural gas into Canada. The government's objective in imposing a tariff can follow one of two opposing principles. The tariff may be used to increase the delivered price of foreign imports to the point where they are effectively shut out of the Canadian market by domestic production. If the tariff is set at the minimum level[2] which prevents American producers from selling gas in Canada, then the tariff equals the difference in the costs of Canadian over American producers. In the case where a tariff eliminates all foreign competition, the government revenue is zero. An alternative government policy, then, may be to maximize tariff revenue. In this case, if the government perceives the demand and supply curves as a monopsonist for foreign gas, it can set the price which maximizes its own revenue.

In 1966, 122 MMCFD of American gas was imported into Canada. The free trade model suggests that all eastern Canadian demand, 754 MMCFD, should be met by American suppliers. It is not *a priori* obvious which policy the Canadian government was following. The large gap between

---

1 / In 1924 the federal government imposed a tariff of six cents per MCF on all imports of natural and manufactured gas. In 1953, the tariff was lowered to three cents per MCF. The tariff was removed as of January 1, 1969 in the Kennedy round of GATT.

2 / Any tariff above this level also shuts out foreign competition.

the actual and the hypothetical volume of imports and the small amount of actual government tariff receipts ($3700 per day) suggests that the policy was one of preventing entry.

In order to test the hypotheses, the model should ideally be rerun for 1956.[3] A solution could be found for a hypothetical free trade model for that year without a tariff. For a second solution, the 3¢ tariff would be imposed on all possible flows from American suppliers to Canadian markets. If the tariff was sufficient to prevent entry, the second solution would show eastern Canadian demand being met by domestic sources. If imports still flowed, the model could be rerun with other tariff charges in order to discover whether the 3¢ tariff maximized government revenue.

This method would have involved extensive data collection. Instead the tariff was added into the framework of the model for 1966. From its impact in 1966, and under realistic assumptions, the effect of the tariff in 1956 can be deduced. The prices used in this experiment are the price set using 'old' contract prices for the American fields.

The 3¢ tariff was an insufficient explanation of the failure of American gas to enter Canada in 1966. The 3¢ tariff prevented no gas from entering rather than the 633 MMCFD *difference between the hypothetical free trade and the actual worlds.*

The tariff necessary to eliminate all American supplies from Canada in 1966 was 11.5¢ per thousand cubic feet.

The trans-Canada pipeline was built in 1956, not 1966. The 3¢ tariff may have been sufficient then. If the 3¢ tariff was pre-emptive in 1956 but an 11.5¢ tariff was prohibitive in 1966, the costs of Canadian suppliers must have risen *relative* to American suppliers by 8.5¢ in the decade 1956 to 1966.[4]

No evidence can be found of this relative increase in Canadian costs. The assumption that the cost disadvantage of Canadian domestic producers in eastern Canadian markets in 1956 was 3.0 cents is therefore untenable. A 3¢ tariff was an insufficient addition to American costs in 1956 to prevent American entry. The building of the trans-Canada pipeline cannot be considered to be a result of the tariff.

An alternative policy, an export tax, is analysed and found to be superior, in terms of benefits to Canadians, to a tariff.

---

3 / The year in which the trans-Canada pipeline was begun.

4 / An absolute increase in all prices which does not change relative costs cannot therefore change comparative advantage or the minimum preventive tariff.

2 THE MODEL FOR THE TARIFF SIMULATIONS

The objective function of the free trade model given in equation 2.8 is altered slightly to incorporate the tariff.

$$\text{minimize } R = \sum_i \sum_j (ad_{ij}g_{ij} + p_i)x_{ij} + \sum_{i=5}^{19} \sum_{j=1}^{9} rx_{ij} \quad (6.1)$$

a, d_{ij}, g_{ij}, p_i, x_{ij} are as given previously; r is the tariff surcharge which is added to all possible shipments from American fields ($i = 5, ..., 19$) to Canadian markets ($j = 1, ..., 9$). The constraints on this objective function are identical to those for the free trade model.

$$\sum_i x_{ij} \geq D_j \quad (j = 1, ..., n), \quad (6.2)$$

$$-\sum_j x_{ij} \geq -S_i \quad (i = 1, ..., m). \quad (6.3)$$

3 THE THREE-CENT TARIFF

The model (6.1–6.3) was first run with a tariff (r) equal to 3¢ per MCF. *No* change in flows results from the addition of a 3¢ tariff. The flows are identical to those of the free trade model. A 3¢ tariff was in 1966 insufficient to prevent the entry of any American gas into Canadian markets.

4 THE PROHIBITIVE TARIFF IN 1966

The 3¢ tariff was insufficient in 1966 to prevent American penetration of Canadian markets. The tariff necessary in 1966 to prevent entry into each Canadian market can be readily estimated.

The minimum preventive tariff has been defined in the first page of this chapter as equal to the cost advantage of the foreign supplier over the Canadian gas producer in domestic markets.

In the chapter on the dual variables, it was suggested that the change in dual delivered prices in eastern Canadian markets between the constrained and free trade solutions ($\bar{u}_j - u_j$) was equal to the cost advantage of American suppliers over Alberta. Therefore, this change in delivered price ($\bar{u}_j - u_j$) represents the minimum tariff necessary to pre-

TABLE 6.1
The prohibitive tariff in 1966[a]

Market	Tariff ($/MCF)
Ontario NW	0.0621
Ontario SW	0.1147
Ontario–Toronto	0.0959
Ontario SE	0.0850
Quebec	0.0886

a / The change in dual values at the markets ($\bar{u}_j - u_j$) between the constrained and free trade solutions.

vent foreign entry into each market.[5] The prohibitive tariff for each eastern Canadian market is equal to the change in the marginal delivered price at that market when Alberta is constrained to meet demand. These preventive tariffs are given in Table 6.1.

The minimum preventive tariff ranges from 6.2¢ per MCF in Ontario Northwest to 11.5¢ per MCF in Ontario Southwest. The tariff which prevents all American gas from entering Canada in 1966 is 11.5¢ per MCF.

It was noted earlier that imports do cross the border, 121 MMCFD in 1966. These imports were purchased almost entirely by Union Gas Company, the distributor in Southwestern Ontario. The daily demand for Ontario SW in 1966 was 187 MMCF. The tariff which leads to a solution which is close to the actual flows is 11.2¢.

The tariffs presented in Table 6.1 are the amounts which remove the last unit of American imports from each market. Lower tariffs might prevent a significant portion of the imports from entering the market. To verify this observation, the tariff model was rerun 15 times, stepping up the tariff by half-cent intervals from 3.0¢ to 11.5¢. Analysing these seventeen solutions will also yield the tariff which maximizes government revenue.

5 THE FOUR-CENT TARIFF

At a tariff level of 4¢ per MCF, American supplies from Kansas to Ontario Northwest are replaced by a west-east trade flow within Canada from Saskatchewan to Ontario NW.

5 / The prohibitive tariff is really some small number above this value. Where the tariff just equals the cost advantage, consumers are indifferent to a choice between the domestic and the foreign supplier.

TABLE 6.2
Change in flows: solution with 4¢ Canadian tariff and free trade solution

Field \ Market	MMCF/day 4 Manitoba	5 Ontario NW	13 US4	16 US7	18 US9
2 Alberta	103				(103)
3 Saskatchewan	(103)	103			
7 Louisiana South			(103)	(103	
9 Texas Permian				(103)	103
15 Kansas		(103)	103		

The tariff increases delivered costs in Canada in two ways. American supplies which continue to penetrate Canadian markets cost more because of the tariff. In addition, where the tariff changes trade flows, shipments appear which are inefficient. They would not appear under free trade. Consumers then bear increased transportation charges and some portion of the tariff surcharge. The tariff revenue paid by the consumer is not a true social cost. Tariff payments are a redistribution of revenue from eastern Canadian gas users to the government. The true social costs, the technological inefficiencies of a tariff are the transportation inefficiencies it engenders.

As was pointed out in chapter 3, one of the limitations of the static linear model is that demand is assumed to be perfectly inelastic. The imposition of a tariff can be considered analogous to a per unit sales tax. In this case of perfectly inelastic demand, the sales tax is passed on completely to the consumer. Due to the assumptions of this model, the tariff is paid entirely by the consumers. The results of the tariff simulations exaggerate the true consumer costs of a tariff.

In the discussions which follow, tariff charges will be included in the cost calculations. It should be remembered that these tariff revenues are merely transfers and not true social costs of policy.

Table 6.2 shows the change in flows when a 4¢ tariff is imposed on all possible flows into Canada from American sources. Table 6.3 gives the increase in the value of the objective function distributed among demand points when free trade is replaced by the 4¢ tariff.

The imposition of the 4¢ tariff affects ten flows including shipments to American markets 4 (Chicago), 7 (Texas), and 9 (California). Two of these three American markets save on transportation costs.[6] The two

6 / This is consistent with the results obtained when Alberta is constrained to meet demand in eastern Canada.

TABLE 6.3
Change in costs at the market: solution with 4¢ Canadian tariff and free trade solution, $/day

Market	Change in transportation costs	Tariff Cost	Tariff Change	Total change in costs
4 Saskatchewan	3490	—	—	3490
5 Ontario NW	7335	1200	1200	8535
6 Ontario SW	—	7480	7480	7480
7 Ontario–Toronto	—	9720	9720	9720
8 Ontario SE	—	3560	3560	3560
9 Quebec	—	3560	3560	3560
13 US4	(4110)	—	—	(4110)
16 US7	(2975)	—	—	(2975)
18 US9	1370	—	—	1370
Total	5110	25,520	25,520	30,630

Canadian markets which experience a change in flows as a result of the imposition of the tariff face higher transportation costs. Inefficient transportation charges in Canada amount to $10,825 per day while tariff revenue is $25,520 per day. American consumers of course pay no tariff costs and save $5715 per day in transportation charges.

6 THE FIVE-CENT TARIFF

Increasing the tariff from 4.0¢ to 5.0¢ per MCF eliminates American suppliers completely from the Ontario Northwest market.[7] The changes in costs on increasing the tariff by 1.0¢ are given below in Table 6.4.

The 1¢ increase in the tariff increases transportation charges in Canada by $4445 while further reducing activity costs in the USA by $2390. The Chicago market again benefits the most by the reduction in American shipments to Canada. Total daily tariff revenue is $30,400 which represents an increase of $4880 over the revenue from the 4.0¢ tariff.

7 / American supplies are totally cut out of the Ontario NW market when the tariff is 5.0¢. However, in Table 6.2, the minimum prohibitive tariff for Ontario was estimated to be over 6.0¢ when dual prices are used. The difference is explained by the techniques. The constrained solution required that Alberta ship to eastern Canada. In the tariff model, any Canadian supplier can ship to eastern Canada. As a result, Ontario NW is supplied by a combination of Alberta and Saskatchewan in the tariff solution. As a balancing item, the costs to Saskatchewan consumers increases in the tariff solution but not in the constrained solution.

TABLE 6.4
Change in costs at the market: solution with 5¢ tariff and solution with 4¢ tariff, $/day

Market	Change in transportation costs	Tariff Cost[a]	Tariff Change[b]	Total change in costs
5 Ontario NW	4445	—	(1200)	3245
6 Ontario SW	—	9350	1870	1870
7 Ontario–Toronto	—	12150	2430	2430
8 Ontario SE	—	4450	890	890
9 Quebec	—	4450	890	890
13 US4	(1715)	—	—	(1715)
16 US7	(1245)	—	—	(1245)
18 US9	570	—	—	570
Total	2055	30400	4880	6935

a / Total tariff revenue.
b / Increase in tariff revenue over 4.0¢ tariff.

7 THE NINE-CENT TARIFF

Further increases in the tariff up to 9.0¢ did not lead to any changes in flows. Table 6.5 gives the changes in costs between the solution with a tariff of 9.0¢ and one using 5.0¢. Transportation costs increase by $17,595 per day in eastern Canada when the tariff is raised to 9.0¢. American consumers are marginally better off by $1095 per day.

Since American suppliers are eliminated from the Ontario SE and Quebec markets, $8900 in tariff costs are eliminated. The increase in activity costs to these two markets of $14,090, however, is greater than the saving in tariff payments.

8 THE TEN-CENT TARIFF

At a tariff level of 10.0¢ per MCF, American suppliers are shut out of the Ontario-Toronto market leaving them shipping only to Ontario SW.[8] The effects of the increase in the tariff to 10.0¢ is shown in Table 6.6.

Although American supplies had been eliminated into Ontario NW with

8 / This is one case where the linear approximation of allowing flows from one source to one market leads to an odd result. A pipeline from Western Canada to the point furthest east (Quebec, 9) would supply Ontario SW (6) also. The tariff results show that markets need different amounts of protection. See Waverman (Sept. 1972).

TABLE 6.5
Change in costs at the market: solution with 9¢ tariff and solution with 5¢ tariff, $/day

Market	Change in transportation costs	Tariff Cost	Tariff Change	Total change in costs
6 Ontario SW	—	16830	7480	7480
7 Ontario–Toronto	3505	17280	5130	8635
8 Ontario SE	8650	—	(4450)	4200
9 Quebec	5440	—	(4450)	990
13 US4	(1000)	—	—	(1000)
16 US7	(2570)	—	—	(2570)
18 US9	2475	—	—	2475
Total	16500	34110	3710	20210

NOTE: With the tariff at 9.0¢ per MCF only two markets remain supplied by American producers – Ontario SW and Toronto.

TABLE 6.6
Change in costs at the market: solution with 10¢ tariff and solution with 9¢ tariff, $/day

Market	Change in transportation costs	Tariff Cost[a]	Tariff Change[b]	Total change in costs
5 Ontario NW	3060	—	—	3060
6 Ontario SW	—	18700	1870	1870
7 Ontario–Toronto	9855	—	(17280)	(7425)
17 US8	(8045)	—	—	(8045)
18 US9	13345	—	—	13345
Total	18215	18700	(15470)	2805

a / Total tariff revenue.
b / increase in tariff revenue over 9.0¢ tariff.

a 5.0¢ tariff, a tariff increase to 10.0¢ further increases the costs for Ontario NW. Saskatchewan, which shipped to Ontario NW at the 5¢ tariff level, ships to Ontario-Toronto with a 10¢ tariff. Alberta, a higher cost shipper, replaces the shipments into market 5.

Although the replacement of foreign suppliers by domestic suppliers increases the real costs of supplying Ontario-Toronto by $9855, the tariff saving of $17,280 makes the market better off in terms of dollar flows.

Total activity costs in the USA increase at this tariff level. This net increase is made up of a large loss to California and a gain to the midwest market.

TABLE 6.7
Change in costs at the market: solution with 11.5¢ tariff and solution with 10¢ tariff, $/day

Market	Change in transportation costs	Tariff		Tota change in costs
		Cost	Change	
6 Ontario sw	20,100	—	(18,700)	1400
7 Ontario–Toronto	3490	—	—	3490
13 US4	(2095)	—	—	(2095)
16 US7	(5405)	—	—	(5405)
17 US8	(7835)	—	—	(7835)
18 US9	13,185	—	—	13,185
Total	21,440		(18,700)	2740

9 THE ELEVEN-AND-A-HALF-CENT TARIFF

The tariff has to be increased to 11.5¢ for all American supplies to be eliminated from the Canadian gas markets. At a tariff level of 11.2¢, the imports of American gas into Canada corresponds to the level and consumption locus of the actual imports for 1966 – 121 MCFD into Ontario SW. The costs imposed by increasing the tariff from 10.0 to 11.5¢ per MCF are shown in Table 6.7.

While total costs change marginally at Ontario SW, the real costs of supplying the market increase by $20,100 per day.[9] The large reduction in the transfer payment (tariff charges) almost offsets this increase.[10] American markets have a reduction of $2150 in the costs of meeting demand. California, however, has a large increase in costs, while three other markets face costs $15,335 lower per day.

10 SUMMARY OF TARIFF CHANGES

In moving from one tariff level to another, as has been done above, the total effect can be lost. Some changes in moving from one level to another can be reversed in further tariff increases. In Tables 6.8 and 6.9, the

9 / At this tariff level, supplies from the Ontario field are used.

10 / The increase in costs at market 7 is due to the replacement of supplies by Alberta for shipments from Saskatchewan.

TABLE 6.8
Difference in primal flows: tariff of 11.5¢ per MCF compared with free trade solution

	Market										
Field	4	5	6	7	8	9	13	16	17	18	Excess capacity
2	103	146	84	200	89	89			(379)	(331)	
3	(103)		103								
4				43							(43)
6			(187)		(89)		276				
7							(422)	422			
9				(243)		(89)		(422)		711	43
14									379	(379)	
15		(146)					146				

TABLE 6.9
Change in costs at the market:
tariff of 11.5¢ per MCF compared with free trade solution

Market	Change in total costs	
4	3490	
5	15,095	
6	20,100	
7	16,550	
8	8650	
9	5440	69325
13	(8915)	
16	(12,195)	
17	(15,880)	
18	31,060	(5930)
		63695

changes in flows and costs are shown for a movement from a zero tariff (free trade) to an 11.5¢ tariff.

The comparison in Tables 6.8 and 6.9 is based on the tariff which eliminates all American suppliers from eastern Canadian markets. Where 121 MMCFD are still allowed to enter the Ontario SW market, the increase in costs for North America as a whole is $50,500 with consumers in eastern Canada paying $52,800 more in the solution to the preventive tariff model.

Both the tariff and constrained models eliminated American suppliers from Canadian markets. The costs to consumers in the tariff model is somewhat lower since other Canadian fields besides Alberta are allowed to ship into eastern Canada.

TABLE 6.10
Canadian government revenue from alternative tariffs

Tariff (¢/MCF)	Imports (MMCF/day)	Total government revenue ($/day)
3.0	754	22,620
4.0	651	25,040
5.0	608	30,400
6.0	608	36,480
7.0	608	42,560
8.0	608	48,640
9.0	379	35,010
10.0	187	18,700
10.5	120	12,600
11.0	120	13,200
11.5	—	—

11 GOVERNMENT REVENUE

Instead of preventing American supplies from entering Canada, the objective of tariff policy in natural gas may have been to maximize government revenue.

Table 6.10 shows the tariff revenue to the Canadian government for tariff levels ranging from 3.0 to 11.5¢ per MCF. The 3¢ tariff is neither a local nor a global maximum. A tariff of 8.0¢ per MCF would have maximized government revenue in 1966. $48,640 per day ($17,700,000 per annum) would be collected at this tariff level.

The 3.0¢ tariff, the actual tariff, in 1966 differs substantially from both the import preventing and the revenue maximizing tariffs.[11]

12 1956, TARIFFS, AND THE BUILDING OF THE CANADIAN EAST-WEST PIPELINE

The trans-Canada pipeline was begun in 1956, not in 1966. Relative costs of production and transportation may have been very different in 1956 from what they were in 1966. The failure of the 3¢ tariff to prevent American suppliers from reaching Canadian markets in 1966 is not

11 / The Canadian government in dropping this tariff in the Kennedy round negotiations in 1968 was giving away very little.

evidence that the same tariff is an insufficient explanation for the building of the trans-Canada pipeline in 1956. The cost disadvantage of Canadian producers may have been only 3¢ in 1956. Once the Canadian pipeline is built, this cost disadvantage may increase as long as demand is less than the capacity of the pipeline. To import American gas into Canada in 1966 involves comparing the investment plus operating costs of a new line from the USA with the operating cost of the trans-Canada.[12]

We know that the necessary unit preventive tariff (the cost advantage of American producers) in 1966 lay between 11.2¢ (the tariff level which allowed entry of 120 MMCFD) and 11.5¢ (the tariff level which prevented all entry). The possible changes in relative costs of the two sets of producers between 1956 and 1966 can be analysed to discover whether a 3¢ cost advantage for American producers in 1956 is realistic.

Changes in comparative advantage could have been caused by any (or all) of the following. (1) Transportation costs may have increased relatively for Canadian suppliers. Or more likely, transportation costs may have decreased for American producers. (2) Production costs may have increased for the Canadian producer relative to the American.

It is difficult to conceive of realistic examples which would lead to a decrease in transportation costs for American suppliers. For this to occur, some technological improvement must be available to Americans in 1966 which was unavailable to them in 1956. This is improbable in this model. The transportation cost data actually used was for 1959. Flows in the model were allowed from any American field to any Canadian demand point. Under these two conditions, the 1966 set of possibilities open to American producers cannot be very different from the set of possible lines which existed in 1956.

The average weighted production costs[13] for all the American fields in the model is 16.11¢. Table 6.11 gives the index of the average well-head price of American natural gas for 1956 to 1966. Using this index, the average production cost in 1956 was 10.98¢. Actual production costs in the USA increased by 5.3¢ from 1956 to 1966 (from 10.98 to 16.11¢).

---

12 / Any profit maximizing firm (perfect competitor or monopolist) will introduce new equipment when the average costs of using the new are less than the average variable costs of the old.

Once the capacity on the existing Canadian pipeline is fully used, the tariff regains its importance. The investment and operating costs of all new additions to capacity must be compared.

13 / Really average FPC area rate price where weights are 1966 production.

TABLE 6.11
Index of average annual wholesale price of gas, USA, 1950–66, ¢/MCF (January 1, 1958 = 100)

Year	Index
1950	55.9
1955	82.4
1956	88.1
1957	97.3
1958	101.7
1960	116.6
1966	129.3

SOURCE: Based on Bureau of Mines estimates of wellhead prices and average prices: American Gas Association, *Gas Data Book* (1967), 22.

To prove that the 3¢ tariff was sufficient in 1956, we must prove that American production costs fell relative to Canadian production costs by 8.2 to 8.5¢ (11.2–11.5¢ less the tariff of 3¢). American costs (per MCF) rose by 5.13¢ in the decade; Canadian costs must therefore have risen by 5.13¢ *plus* 8.2 to 8.5¢ for the story to be true. Assuming a prohibitive 3-¢ tariff in 1966, we infer that production costs must have risen by 13.33 to 13.63¢ in Canada between 1956 and 1966. But in 1966 average production cost in Canada was 12.35¢ per MCF. Deducting our inferred cost rise in Canada from this figure leaves a negative inferred production cost estimate for 1956. This 1956 production cost estimate is an improbable value. Production costs in 1956 must have been greater than this.

It is therefore safe to conclude that the actual cost difference between American and Canadian suppliers in eastern Canadian markets was greater than 3¢ in 1956. The 3¢ tariff could not therefore have been prohibitive in 1956. Its presence is an insufficient explanation of the Canadian east-west pipeline.

13 BALANCE OF PAYMENTS

The free trade and the constrained solutions present very different international trade flows. Under very general assumptions, the effects on the Canadian balance of payments of both solutions can be compared.

In the free trade model Canada exports 1914 MMCFD to the USA and imports 633 MMCFD from that country. In the constrained model

(Alberta is constrained to meet eastern Canadian demand) Canada ships 1281 MMCFD to the USA. In this solution, the imports into Canada are zero. It is not *a priori* obvious that the reduction in the *value* of the exports will exactly balance the reduction of *value* of the imports.

The question arises of how to value the flows. What price is placed on the Canadian earnings of a unit of Canadian exports? What percentage of the delivered price of American gas at Toronto is earned by Americans? Two different approaches were used. First, the shipments were valued at the marginal delivered price (dual price) at the market supplied. The market dual is the sum of three components, the production costs, the transportation costs, and the field rental royalty. Foreigners, under this assumption, own all the facilities for their exports from the well to the city gate. In this case, the net foreign exchange earnings of Canada in the free trade solution are $256,800 per day. The addition of the Canadian east-west flow constraint increases these daily net foreign earnings to $284,400, an increase of $27,600.

In the second approach, the international flows were valued at the sum of production costs and field rentals. This approach does not impute any transportation earnings to the flows. In the free trade model, Canada nets $145,450 in foreign exchange. The existence of the Canadian east-west trade increases these earnings to $174,800.

Depending on which assumption is made as to the ownership of transportation facilities, the introduction of a Canadian transcontinental flow increases daily foreign exchange earnings by $27,600 to $29,400. The higher figure arises from the case where *no* transportation charges are imputed to the producer.

The two estimates represent extreme cases. The real change in foreign exchange earnings will be somewhere between the two. The Canadian markets served by American fields are close to the Canadian American border. Assume that the transportation link in a country is owned by the nationals of that country. On American imports into Canada, we assume that the transportation earnings are by Americans. Of the three American markets for Canadian gas, two (US districts 8 and 10) are near the Canadian border while one (California) is approximately 800 miles away. Under the above assumption of ownership, the transportation earnings on Canadian exports to the USA are divided nearly equally between Canadians and Americans. The ratio of transportation costs on imports into Canada to the transport costs on exports out of Canada is four to one. Canadians, it is assumed, therefore receive none of the transport receipts on imports into Canada and one-half of the transportation earnings on exports from Canada. Canada then receives one-ninth (1/9) of the earnings from transporting gas. A good estimate of the increase in the surplus on balance of payments account due to the inclusion of a Canadian transcontinental flow is $28,800 per day.

This increase[14] in the balance of payments surplus represents neither an increase in Canadian gross national product nor an increase in welfare. Canada is using gas which could be exported to replace imports of the fuel. To argue that this is a real gain is to argue that the elimination of imports into Canada of all products in 1966 by redirecting exports to the domestic economy would have increased GNP. When imports are replaced from domestic supplies because of an artificial barrier, GNP and welfare are reduced. Real costs in Canada, as was noted before, *increase* by some $50,800 to $54,100 per day due to the prevention of American imports into Canada. This, not the change in the balance of payments, is the real impact of the barrier.

To add further weight to the argument that this section should be ignored is the fact that the figure used, $28,800, exaggerates the gain in the balance of payments.

$28,800 is the *direct* gain to Canada. There are *indirect* costs to this import substitution. The cost of natural gas increases to users in Canada. In the short run, supply will be decreased by these other producers. In the long run if they cannot find a cheaper source of fuel than natural gas, supply of gas intensive goods will still be below the amount which would be produced without the Canadian line. To the extent that these producers are exporters of goods which have an elastic demand, foreign exchange earnings will fall.

14 AN EXPORT TAX[15]

Besides using the two instruments of the tariff and the exportable surplus policy, the government could have alternatively introduced an export tax.

---

14 / $28,800 a day is $10,500,000 per year. The balance on *commodity* account for Canada in 1966 was $458,880,000. $10,500,000 is less than 3 per cent of this balance. The figure of $10,500,000 represents three-fifths of one per cent of the Canadian deficit on current account with the USA.

15 / M. Wolfson of Oregon State University gave me this idea during a seminar at the University of Durham.

A tax on all units of gas leaving Canada would have the same effect as the tariff or the constraint – the elimination of American supplies into eastern Canada. The tariff accomplished this objective by raising the delivered price in eastern Canadian markets to induce Alberta to ship through. The export tax raises the price of western Canadian gas in American markets to the point where Canadian producers are shut out of these markets and ship instead into eastern Canada.

Table 6.12 gives the level of exports from Canada under different export tax levels. The revenue accruing to the Canadian government is shown[16] as are the west-east shipments in Canada, and the excess capacity in Canadian fields.

At an export tax level of approximately 12¢ per MCF, all American supplies are shut out of eastern Canadian markets. While both BC and Alberta have excess capacity, the revenue earned by the Canadian government is nearly $100,000 per day. Export taxes of either 8.0¢ or 11.0¢ per MCF would maximize government revenue at $123,000 per day (nearly $500,000,000 on a present value, discounted at 10 per cent).

On pure Canadian interests, the export tax is clearly preferable to a tariff. While achieving the same effect, the export tax earns $100,000 per day from Americans compared to the tariff's zero revenue (assuming 100 per cent forward shifting). The change in delivered prices in eastern Canadian markets in these models is greater with a tariff than with an export tax. The reason is clear. In both cases, shadow delivered prices must rise by a large enough amount in eastern Canada to offset the cost disadvantage of western Canadian producers. With the tariff however, prices in eastern Canada must rise by an additional amount to cover the increase in royalty value at Alberta. The export tax produces excess capacity in Alberta, hence the royalty value must decrease from a positive value in the free trade solution to zero in the export tax solution.

Maximizing the return to Canada with the constraint that eastern Canada be supplied by western Canada is achieved with an export tax. The desired west-east shipments occur. The government revenue is sufficient to compensate eastern Canadian consumers so that they pay no more for their gas than if Americans supplied these markets.

Western Canadian producers are worse off in two respects when trade

16 / The export tax is assumed to be shifted completely on to American consumers.

TABLE 6.12
Export tax on shipments of Canadian gas into American markets

Tax	Exports from Canada (MMCFD)	Tax revenue ($/day)	Change in Canadian west-east shipments	Excess capacity, Canadian fields (MMCFD)	American markets lost
1.0	1915	19,150	none	none	none
2.0	1832	36,640	none	BC – 83	Part California
3.0	1832	54,960	none	–	–
4.0	1729	69,160	Sask. to Ontario NW	–	–
5.0	1583	79,150	none	BC – 222 Al – 7	Part Pacific NW All California
6.0	1583	94,980	none	–	–
7.0	1536	107,520	none	BC – 222 Al – 11	–
8.0	1536	122,880	none	–	–
9.0	1358	104,220	Alberta to Ont. SE & Quebec	–	Part Pacific NW, Midwest All California
10.0	1115	111,500	Alberta to Toronto	–	–
11.0	1115	122,650	none	–	–
12.0	928	99,360	Alberta to Ont. SW	–	–

NOTES: BC = British Columbia; Al = Alberta; dash denotes no change from previous row is limited through an export tax. The producers lose $32,850 per day because the shadow royalty value of their production drops to zero. They also lose $24,380 per day since some of the production is unused in the export tax solution. However, the government revenue from the 11.0¢ export tax is sufficient to compensate producers in western Canada for their losses, as well as consumers in eastern Canada.

Under the assumptions that an export tax is shifted forward 100 per cent and that Americans do not retaliate, Canadians could achieve their National Policy by taxing Americans to pay for the ensuing losses.

Whether this policy is realistic depends on the realism of the assumptions. In all likelihood, the tax would not be completely shifted and retaliation would certainly be possible.

One other interesting and important point emerges from the export tax discussion. An export tax up to 3.0¢ per MCF would have no trade diversionary effects. It would then appear that Canadian producers are not taking full advantage of their locational advantages in western American markets. Prices could be raised by 3.0¢ per MCF, thus increasing revenue by $20,000,000 per annum.[17]

---

17 / If a profit maximizing monopolist were in charge of western Canadian production, he would raise export prices by 8¢. Revenue would increase by $123,000 per day, costs (implicit) by $57,200 (losses in value of production and excess capacity), and profits by $65,800 per day.

7

Sensitivity of the results – alternative data

1 INTRODUCTION

Throughout this book, the 'reasonableness' of the model has been emphasized. The flow pattern of the constrained solution is very close to the actual flows of 1966.

While the model may seem to give a reasonably accurate portrayal, its interpretations would be misleading if the results depended on the specific data used. The estimates of costs were by no means perfectly accurate. If slight changes in cost estimates led to major changes in conclusions, then the model is not at all interesting. If adjustments to demand or supply change important results, the model is not a useful tool for policy analysis.

In this chapter, the sensitivity of the main findings to the underlying data is tested. The first necessary step is to determine the criteria by which sensitivity should be measured. A number of the supply points are quite close together. Therefore, I do not consider the model sensitive if small changes in underlying cost data (say 10 per cent) shift a flow to a market from one field to a nearby field. If, however, a 10 per cent change in costs changes the source of supply to a Canadian market from an American to a Canadian field, then the model is sensitive.

The results are sensitive to the underlying data if significant changes in costs alter the major conclusions of the study. A number of methods were used to test the sensitivity of the results.

2 CHANGES IN COST DATA

Changes in the costs of positive flows

The first method used was to increase the costs of each positive flow in the primal free trade solutions. This then increases the costs of flows from the USA to Canada while it leave the costs of flows within Canada unchanged.[1] This increase in relative costs for the positive flows led to the replacement of 22 (two-thirds) of them. While a large number of flows do change where relative costs are changed by 10 per cent, this is *not* an indicator of sensitivity. All of the replacements involved a shift from one field to a nearby field. Specifically, the flows from American fields to Canadian markets would be replaced by shipments from other American fields to these Canadian markets. No flow along north-south lines was replaced by an east-west flow.

Transportation costs

As a second method of determining the dependence of the findings on the specific data used, alternative values of transportation costs were used. Lowering transportation costs should tend to improve the ability of western Canadian fields to compete in eastern Canadian markets since a major disadvantage facing western Canadian producers is the distance to these markets. The estimate of transportation costs (*a*) actually used was 1.1¢ per MCF per hundred miles. Transportation costs were lowered to 0.5¢ (a decrease of over 50 per cent) and raised to 2.3¢ (an increase of over 100 per cent). Neither change led to major differences in trade movements. In no case did a transcontinental east-west flow appear.

Table 7.1 gives the flows in the free trade model where the marginal costs of transportation (*a*) has been lowered to 0.5¢ per MCF per 100 miles (and the old price series has been used to represent production costs in the USA). Comparing this table to Table 4.1, where identical data are used but the marginal transportation charge is 1.1¢ per MCF per 100 miles, shows *no difference* in transborder flows. When marginal transportation costs are halved the only change is the transference of excess capacity from fields in Mexico and Louisiana south (5, 7) to California

1 / Increasing the costs of a basic activity to replace it by a non-basic activity.

TABLE 7.1
Free trade model: adjusted transportation cost ($a = 0.50$)

From \ To		BC 1	Alta 2	Sask. 3	Man. 4	Ont. NW 5	Ont. SW 6	Ont.–Tor. 7	Ont. SE 8	Quebec 9	US1 10	US2 11	US3 12	US4 13	US5 14	US6 15	US7 16	US8 17	US9 18	US10 19	Slack
British Columbia	1	219																		222	
Alberta	2		572	176														979	332	382	
Saskatchewan	3				103																
Ontario	4																				43
Mexico	5																150				
Louisiana North	6						187		89					1391							
Louisiana South	7											7960	2596	1189			510				
Texas Gulf	8																10,662				
Texas Permian	9							243		89							2212		1006		
Texas Midcontinent	10																1790				
Texas Panhandle	11															3250					
San Juan	12																		1397		
New Mexico SE	13																		1338		
Wyoming	14																		667		
Kansas	15					146								198	1978						
Oklahoma	16													3106	596						
Mississippi	17												429								
California	18																		1656		233
West Virginia	19										524	56									

(18).[2] The reduction in transportation costs does not generate transcontinental east-west flows.

Production costs

The estimates of production costs used in this study are *prices* at the fields, not estimates of true production costs. At least for American fields engaged in inter-state commerce, the use of prices for costs is not a serious error since these prices (area rates) are regulated according to costs of production and normal rates of returns.

Using the prices received by unregulated Canadian fields as a measure of their production costs will lead to error if the average profit at these Canadian fields is above that for the American fields. If this profit difference exists (and the interpretation of the dual variables would suggest that it does), then the prices used to represent production costs in Canada contain royalty payments. These royalties should be deducted from the prices.

No attempt was made to measure the relative profitability of American and Canadian fields. Instead, the prices used for Canadian field costs were reduced by 20 per cent to represent royalties earned. This twenty-per-cent reduction surely exaggerates both the absolute and relative profitability of Canadian fields. However, using these adjusted prices will show the model's sensitivity to the relative production cost estimates.

Several other changes were made in the estimates of production costs. The prices at the fields of California, Ontario, and West Virginia were lowered. These three fields are very close to the markets and it was felt that the prices already used contained large elements of location royalties.[3]

The adjusted production costs are given below:

	$/MCF
1 British Columbia	0.0845
2 Alberta	0.1070

2 / The reduction in transportation costs makes several fields cheaper in the Californian market than the Californian field. This indicates that the production costs assumed for California are too high. In the next section, production costs are revised.

3 / Ontario production was always excess in the estimates. Yet, it clearly is used. The lowering of transportation costs in the previous section led to excess capacity at California.

3 Saskatchewan	0.1324
4 Ontario	0.1540
5–17	FPC 'old prices'
18	0.1428
19	0.1302

Table 7.2 gives the free trade flows using these adjusted production costs (all other data the same as the original model). Again, the results are nearly identical to those of Table 4.1, the basic difference being the shifting of excess capacity away from the Ontario field. As in Table 4.1, western Canadian fields do not ship into eastern Canadian markets.

Use of the dual

The question of determining the sensitivity of the result of the free trade model (no east-west flows) can be posed in another way. What changes in the relative prices at fields are necessary to replicate in a free trade model, the results of the constrained model? The primal solution to the constrained model gave the flows in North America resulting from a barrier to imports into Canada (quota, or other non-market barrier). The dual to this constrained problem establishes the shadow delivered prices in eastern Canadian markets, when these markets must purchase from western Canada. Imports into Canada would approach zero under free trade, if the government set delivered prices in eastern Canadian markets at their values in the dual to the constrained model.

The question of sensitivity is identical to the problem of estimating the minimum preventive tariff discussed in the last chapter. Canadian east-west flows would appear if the relative costs of Canadian producers changed so as to eliminate their cost disadvantage in domestic markets. These cost disadvantages were given on page 86. As a percentage of the costs of shipping from Alberta to eastern Canadian markets, this cost disadvantage ranges from 18.2 per cent in Ontario northwest to 30.6 per cent in Ontario southwest.

As a percentage of the costs of production assumed for Alberta (13.0 cents), the measure of Alberta's disadvantage in eastern Canadian markets ranges from 47 to 86 per cent! While the cost data used were not perfectly accurate, the relative costs of producing in Alberta would have to be overestimated by 86 per cent for the entire east-west Canadian transcontinental pipeline to appear in a free trade model.

TABLE 7.2
Free trade: 1966 adjusted prices

From \ To		BC 1	Alta 2	Sask. 3	Man. 4	Ont. NW 5	Ont. SW 6	Ont.–Tor. 7	Ont. SE 8	Quebec 9	US1 10	US2 11	US3 12	US4 13	US5 14	US6 15	US7 16	US8 17	US9 18	US10 19	Slack
British Columbia	1	219																		222	
Alberta	2		572	176														979	332	382	
Saskatchewan	3				103																
Ontario	4								43												
Mexico	5																				150
Louisiana North	6						187		46					1434							
Louisiana South	7											7960	2596	1146			553				
Texas Gulf	8																10,662				
Texas Permian	9							243		89							2319		773		126
Texas Midcontinent	10																1790				
Texas Panhandle	11															3250					
San Juan	12																		1397		
New Mexico SE	13																		1338		
Wyoming	14																		667		
Kansas	15					146								198	1978						
Oklahoma	16													3106		596					
Mississippi	17												429								
California	18																		1889		
West Virginia	19										524	56									

3 DEMAND AND SUPPLY

Besides the estimates on the cost side, the basic results may depend on the estimates of demand and capacity. In particular, the capacity of a field was assumed to be that field's net production in 1966. The production estimate is itself heavily dependent on the flows created by the trade barriers. Perhaps changing the supply or the demand estimate would therefore lead to a different conclusion as to the relative viability of east-west trade flows.

In chapter 3, page 38, alternative estimates of capacity for 1966 were presented. These estimates were derived by finding the average reserve to production ratio for North America in 1966 and assuming that each field produced at this average ratio. This adjustment increased the estimated capacity of Canadian fields and reduced that of American fields. Canadian producers are prevented from exporting all they desire to American markets and constrained to maintain an inventory in the ground for future Canadian demand. It is likely that the reserve-to-production ratio in Canada has been raised above what would prevail under *laissez-faire*. This adjustment which increases Canadian capacity (lowers the reserve production ratio) is purely artificial. It is probably, however, in the direction which increased trade would lead.

The changes in capacity using this adjustment are drastic. Estimated supply increases by 210 per cent in British Columbia and 220 per cent in Alberta, while falling 22 per cent in Texas Gulf and 50 per cent in Louisiana north.

Table 7.3 gives the flow pattern under the assumption of free trade using this hypothetical supply pattern. Under free trade, western Canada ships through to Ontario Northwest, but no further east.

In Table 7.4, the changes in costs between constrained and free trade models is given for each affected demand point. Costs for North America as a whole increase by only $15,800 per day when trade is constrained (as compared to a difference in the value of the objective functions of $50,500 when actual 1966 production data is used). Costs in Canada increase by $38,300 while the USA gains $22,500 when Alberta is forced to meet eastern Canadian demand.

The great change in estimated capacity between actual production and the hypothetical values has two main effects. The first is to create a free trade shipment from Alberta to Ontario Northwest. As a result, the

TABLE 7.3
Free trade model: 1966 adjusted supplies, MMCFD

From \ To		BC 1	Alta 2	Sask. 3	Man. 4	Ont. NW 5	Ont. SW 6	Ont.–Tor. 7	Ont. SE 8	Quebec 9	US1 10	US2 11	US3 12	US4 13	US5 14	US6 15	US7 16	US8 17	US9 18	US10 19	Slack
British Columbia	1	219																		450	250
Alberta	2		572	176	103	146								105	1978			979	1349	153	
Saskatchewan	3										112										
Ontario	4																				31
Mexico	5																150				
Louisiana North	6						187		89								559				
Louisiana South	7										7	8016	2704				1929				
Texas Gulf	8																7444				
Texas Permian	9																3440				
Texas Midcontinent	10																1744				
Texas Panhandle	11															2766					
San Juan	12																		1564		
New Mexico SE	13							243		89			364				158				
Wyoming	14																		612		
Kansas	15													3186							
Oklahoma	16													2229		1080					
Mississippi	17												321								
California	18																		2871		
West Virginia	19										405										

TABLE 7.4
Changes in costs at affected demand points when Alberta is constrained to ship to eastern Canada: hypothetical adjusted 1966 shipments

Market	Change in costs	
	$	%
4 Manitoba	(2297)	(1.0)
6 Ontario southwest	8382	19.3
7 Ontario–Toronto	17739	23.9
8 Ontario southeast	8651	33.4
9 Quebec	5847	19.7
Subtotal Canada	38322	
10 USA1	(5242)	
13 USA4	12053	
14 USA5	(27067)	
18 USA9	(2254)	
Subtotal USA	(22510)	
Total	15812	

estimated costs to Canada of preventing free trade falls. The increase in capacity at Alberta is shipped mainly to American markets which are close by. As a second result, then, constraining Alberta to ship to eastern Canada makes a number of American markets worse off.

In order to test the model further, and also to estimate the costs of restrained trade were the present restrictions to continue, the models are rerun using values of supply and demand for 1980. Table 3.4 gives estimated values for demand in 1980. Capacity for each field in 1980 is given in Table 3.2. This method of estimating capacity in 1980 is highly unreal. I would be surprised if these figures bear resemblance to actual production in 1980. The exercise is useful, for it indicates the magnitude of costs which the consumer will bear if restrictions remain in force and if fields develop in the same pattern in the future as they have in the past.

Tables 7.5 and 7.6 give the flows in the year 1980, when the supply and demand estimates are used along with the new price series. Under free trade, half of the demand in Ontario Northwest and all other eastern Canadian markets is served by American fields.[4] In the constrained solution, it is assumed that two-thirds of the demand in Ontario Southwest is met by imports (the 1966 value).

In Table 7.7, the change in costs between constrained and free trade

4 / There is also a small positive shipment from field 13 to market 7. See chapter 4, n. 1.

TABLE 7.5
Free trade model, 1980

From \ To		BC 1	Alta 2	Sask. 3	Man. 4	Ont. NW 5	Ont. SW 6	Ont.–Tor. 7	Ont. SE 8	Quebec 9	US1 10	US2 11	US3 12	US4 13	US5 14	US6 15	US7 16	US8 17	US9 18	US10 19	Slack
British Columbia	1	545																		1055	
Alberta	2		967	430	304										3319			2000	2110	445	
Saskatchewan	3					192															
Ontario	4										23										29
Mexico	5																150				
Louisiana North	6						510										940				
Louisiana South	7											12,800	4760				4090				
Texas Gulf	8																12,760				
Texas Permian	9																5750				
Texas Midcontinent	10																3000				
Texas Panhandle	11													55		4695					
San Juan	12																		2680		
New Mexico SE	13									260							1210				
Wyoming	14																		1050		
Kansas	15					203					247			4665	346						
Oklahoma	16							660	240					4800							
Mississippi	17												550								
California	18																		4940		
West Virginia	19										700										

TABLE 7.6
Constrained model, 1980

From \ To		BC 1	Alta 2	Sask. 3	Man. 4	Ont. NW 5	Ont. SW 6	Ont.–Tor. 7	Ont. SE 8	Quebec 9	US1 10	US2 11	US3 12	US4 13	US5 14	US6 15	US7 16	US8 17	US9 18	US10 19	Slack
British Columbia	1	545																		1055	
Alberta	2		967	430	112	395	160	660	240	260					1796			2000	2110	445	
Saskatchewan	3				192																
Ontario	4										23										30
Mexico	5																150				
Louisiana North	6						350										1100				
Louisiana South	7										247	12,800	4760				3843				
Texas Gulf	8																12,760				
Texas Permian	9																5750				
Texas Midcontinent	10																3000				
Texas Panhandle	11													55		4695					
San Juan	12																		2680		
New Mexico SE	13													173			1297				
Wyoming	14																		1050		
Kansas	15													3592	1869						
Oklahoma	16													5700							
Mississippi	17												550								
California	18																		4940		
West Virginia	19										700										

TABLE 7.7
Changes in costs at affected demand points when Alberta is constrained to ship to eastern Canada. 1980 supply and demand, new price series

Market	Change in costs $	%
5 Ontario Northwest	9155	15.2
6 Ontario Southwest	16,176	37.0
7 Ontario–Toronto	52,866	27.3
8 Ontario Southeast	18,528	24.8
9 Quebec	11,856	12.9
Subtotal Canada	108,581	
10 USA1	7015	
13 USA4	(5250)	
14 USA5	(42,516)	
16 USA7	(6860)	
Subtotal USA	47,611	
Total	60,970	

solutions is shown for all demand points where flows shift. While North America as a whole pays little more per day in 1980 than in 1966 due to inefficient transportation patterns ($60,600 as compared to $50,000), the USA is now better off by some $47,600 per day. Eastern Canadian consumers pay $108,000 more per day for gas in the constrained solution or $40,000,000 per year. American consumers on the whole save $17,000,000 per year as compared to what they would pay under free trade.

If present restrictions continue and field capacity develops in the same pattern as it exhibited in 1966, the burden on eastern Canadian consumers will grow.

4 SUMMARY

What have these various exercises shown? The basic results given in chapter 4 relied on arbitrary estimates of costs and capacity. In this chapter, analysing the dual variables (section 2), showed that the relative production costs of western Canadian producers would have to be overstated by some 86 per cent for the actual east-west Canadian flows to fully appear under free trade. Lowering Canadian production costs to take account of the royalty possibly present in the price data had no impact (section 2). Lowering the estimate of the marginal transportation cost by over 50 per cent (section 2) had little impact.

Similarly, adjusting the values of production for 1966 (section 3) to greatly increase western Canadian production had some impact on east-west flows, but did not affect the major conclusions.

Taking hypothetical estimates of demand and supply for some future date (section 3) indicated that the absence of east-west trade flows under free trade was not dependent on the actual year chosen.

On the basis of all this evidence, the conclusion is clear, the absence of east-west flows under free trade is a basic result not dependent on the specific data used in chapter 4.

Conclusions

This study set out to estimate the costs to consumers and the benefits to producers of the governmental policies which restrict trade in natural gas between Canada and the United States.

Demand was taken to be net consumption in 1966. The capacity of fields was assumed to be their net production in 1966. Production costs were estimated by the prices at the fields. For the USA, these prices were regulated by the government. No such price ceilings existed in Canada.

Using these values, the model suggested that the presence of constraints on trade has added $184,000,000 (present value) in inefficient transportation costs to the consumer's bill. The division of this cost between Canada and the USA depended on the specific estimate of production costs in American fields. Under certain assumptions, the restrictions on trade lowered costs to American consumers. In these cases the costs to Canadian consumers rose above $184,000,000.

In Chapter 7, various shocks were introduced into the models by way of alternative data to test the sensitivity of both the basic results and the estimate of the total inefficiency costs. These changes indicated that the basic conclusion was very robust – in the absence of trade restrictions, eastern Canadian demand would be met by American suppliers. Using estimates of demand and production in 1980 showed that as demand and supply grow, eastern Canadian consumers bear a higher proportion of the total costs.

Considering the grossness of the data, I would suggest that the present value of the additional costs to eastern Canadian consumers, of having western Canada as a supply source, lies in the range of $150,000,000 to $200,000,000 (including increased pipeline capital costs). The USA is marginally affected in the range of $20,000,000 (thus a benefit) to $40,000,000. These measures of the additional costs borne by consumers are a lower bound to the true costs. Any attempt to improve the model by reducing its limitations – static, single year, inelastic demand and supply – will only increase the estimates of the inefficiencies created by trade restrictions. This statement can be easily verified. If free trade has any effect in the simple model, then Canadian suppliers to eastern Canadian markets will be replaced by American suppliers. This replacement will occur only if the costs at the markets fall. The Canadian production so freed will move into American markets also leading to a price fall.

The complete model would allow demand and supply to vary. If demand is not perfectly price inelastic, then the price fall in certain Canadian and American markets will increase the quantity demanded. Trade restrictions, because they artificially raise costs, diminish the quantity demanded. Thus, welfare is reduced by the consumers' surplus forgone on the decrease in quantity consumed. On the supply side the changes in prices at supply points will change the relative profitability of supplying areas. As a result, exploration activity will increase relatively at the newly profitable supply points. In the complete model supply would tend to shift away from those fields where production was artificially increased because of trade restrictions.

In the complete model where demand and supply are allowed to vary, consumers will be better off for two reasons – the consumers' surplus gained on the increased quantity demanded and decreases in costs as the more efficient fields are developed.

Having read this far, many politicians may shrug and suggest that two hundred million dollars is not a great amount of money, especially when the benefits of these policies are considered. After all, the point of the study is that the trade pattern in natural gas would be very different without government intervention. Policies were not superfluous. To maintain east-west flows within Canada, restrictions have to be placed on north-south flows.

The question is, why worry about the direction of flows in natural gas? Why is natural gas such an important commodity to the nation?

The number of inefficiencies which we as a nation can bear and survive is limited. Deviations from free market behaviour should be restricted to those cases where it is in the social interests. In reading the pipeline debates (Trans Canada Pipeline, 1956, and Great Lakes Transmission, 1965), one is struck by the similarity in the arguments with those preceding the construction of the transcontinental railroads in the 1870s. Two decisions were involved in deciding on the trans-Canada pipeline. One involved the need for east-west flows. The second involved the necessity of having the route wholly within Canada. Just as in the 1870s, the decision was to remove the potential American influence as far as possible.

The decision to build north of the border is difficult to rationalize. Moreover, Canadian policy is completely inconsistent. In 1952, four years earlier, an oil pipeline was built *through the* USA to link western and eastern Canada. In 1956, the same suggestion is rejected for the gas pipeline. The view that west-east flows of natural gas are important to Canada rests on two grounds. First, that gas pipelines, like railroads, are an important element in social welfare. Second, that this type of self-sufficiency minimizes American influence. Natural gas pipelines are *not* railroads. Natural gas is a homogeneous commodity. One unit from Texas is very similar to one unit from Alberta. Gas is not at all a commodity with cultural overtones. The device which brings natural gas to the home from the field is incapable of any other activity. Unlike railroads, pipelines are not important movers of goods and people between east and west.

It is naive to feel that the mere construction of an 'all-Canadian' pipeline was a sufficient policy to minimize American influence. The natural gas itself is largely owned by American companies. The pipeline itself was built by Americans under American control and using mainly imported labour and materials. The all-Canadian pipeline is a myth. While the pipeline is now controlled by Canadians, other segments of the industry are not. Does it really matter whether we get American gas from Texas or American gas from Alberta?

Why have east-west pipelines at all? Why not just trade, letting markets purchase gas or oil from the cheapest source? The argument against free trade revolve around two main notions – eastern Canadian markets would not be served, western Canadian fields would be quickly depleted.

The results of this book would indicate that in the absence of trade restrictions, eastern Canadian markets would be supplied from American fields. The costs in these markets would be lowered below the present level. I cannot therefore believe that it was necessary to induce western Canadian producers to ship east, for otherwise, demand in eastern Canada would have not been met. Both the lower estimated costs from American sources and the continued pressure for imports from the USA are my evidence.

Several individuals have argued that the limit on exports of gas is necessary in order to prevent the too rapid depletion of these resources.[1] However, restraining the *direction of shipments* is an inefficient means of preventing depletion. Constraints on the *rate of production* would be the easiest means of reducing the rate of depletion.

I can see two other quite rational arguments which suggest that trade, at least on its present terms, should be minimized. The first of these has been analyzed earlier. Canadians are not taking advantage of their quasi-monopolistic advantage in American markets. The price Canadian producers receive is too low. In chapter 6, it was indicated that prices could be raised by 3¢ per MCF without affecting trade (with 1966 data). Prices could be raised by 60 to 70 per cent and increase the revenue received by Canadian producers (even though some trade dislocations occurred). The Canadian government could directly receive this increased revenue by levying appropriate export taxes. The evidence of export prices being set too low does not suggest the prevention of trade. It does suggest that harder bargaining should take place.

There is one more argument (many will see arguments I have omitted) which needs explanation. Let me call this argument 'the paradox of trade.' If eastern Canadian markets contract with American suppliers, these supplies may be cut off in times of emergency. In exporting gas to the USA, long-term commitments are made so that this gas could not be used in Canada, if it was ever needed.

The 'paradox of trade' rests on trading with someone who has power. If Canada imports from the USA, Canada can be hurt because the tap can be shut off south of the border. If the USA imports from Canada, Canada can be hurt because the tap can not be shut off north of the border. The Americans have power – military – to control the taps. Any trade, export or import, makes Americans dependent on, or at least involved in, Canada, and then potentially able to control. The USA could always impose its will on Canada and would do so to improve US welfare, disregarding Canadian welfare.

This view of power is logical and, given present events (the American

1 / See Laxer (1970), 4, 5, for example.

10 per cent import surcharge of late 1971), rational. The trans-Canada pipeline is then a form of insurance policy. Canadians prefer to pay a premium in terms of higher transportation costs to know with certainty that this flow of gas will continue in the future.

While I agree that this is a rational *argument*, the question of whether it is a logical *policy* depends on the premium paid and its relation to the possible damage.

An insurance policy with a premium of $200,000,000 is expensive. Consider the case where free trade is allowed. What is the probability that the USA would use its power to hurt Canadians, by turning off the natural gas tap. Take an extreme probability – 5 per cent. Most Canadians would agree that it is unlikely that the chances are 1 in 20 of both an emergency occurring in the USA and the USA shutting off gas flows to Canada.

Assume that some insurance company was willing to insure against American intervention, the premium being $200,000,000 (present value sum of all future premiums). It would be a rational policy to pay the premium where the chance of collection is 1 in 20, only if the expected costs to Canadians of American intervention were at least $4,000,000,000 ($200,000,000 divided by 1/20).

With a probability of 5 per cent of Americans using their power to decrease Canadian welfare, an insurance premium of $200,000,000 is rational only if the decrease in Canadian welfare resulting from the action is over 4 billion dollars.

At a probability of 1 per cent that the taps would be turned off, costs must be $20 billion to make the premium justifiable. At a probability of 1 in 1000 of such an American action, necessary costs rise to $200 billion!!

I have not estimated what the costs would be to Canada if Canadians engaged in trade in gas but were 'taken' by Americans. Given that I view the probability of such actions as closer to 1 in 1000 than 1 in 20, I doubt if the premium Canadians pay – the costs of an inefficient transportation network – are worth paying.

References

BOOKS AND ARTICLES

ADELMAN, M.A. The supply and price of natural gas. Supplement to *Journal of Industrial Economics*, 1962

— Efficiency in resources use in crude petroleum. *Southern Economic Journal*, 31, 2, Oct. 1964, 101–39

AITKEN, H.G.J. The midwestern case: Canadian gas and the FPC. *Canadian Journal of Economics and Political Science*, 25, 2, May 1959, 129–43

ARROW, K.J. *Social choice and individual values*. Cowles Foundation, Monograph 12, 2nd ed. New York: Wiley, 1963

— Tullock and an existence theorem. *Public Choice*, 6, spring 1969, 105–12

BARNETT, H.J. and CHANDLER MORSE. *Scarcity and growth, the economics of natural resource availability*. Baltimore: Johns Hopkins Press for Resources for the Future, Inc., 1963

BAUMOL, W.J. and R.C. BUSHNELL. Error produced by linearization in mathematical programming. *Econometrica,* 35, 3–4, July, Oct. 1967, 447–71

BUCHANAN, J.M. and G. TULLOCK. *The calculus of consent*. Ann Arbor: University of Michigan Press, 1962

CHENERY, H.B. Process and production functions from engineering data. In W. Leontief, ed., *Studies in the structure of the American economy*. Oxford: Oxford University Press, 1953

DANTZIG, G. *Linear programming and extensions.* Princeton: Princeton University Press, 1963

DORFMAN, R., P.A. SAMUELSON and R.M. SOLOW. *Linear programming and economic analysis.* New York: McGraw-Hill, 1958

DUNN, C.L. The economics of gas transmission. INGAA Conference, Oct. 1954, mimeo

EASTERBROOK, W.T. and H.G.J. AITKEN. *Canadian economic history.* Toronto: Macmillan, 1958

FOGEL, R.W. *The Union Pacific railroad: a case in premature enterprise.* Baltimore: 1960.

FOWKE, V.C. The National Policy – old and new. *Canadian Journal of Economics and Political Science,* 18, 3, Aug. 1952, 271–86

GERWIG, R.W. Natural gas production, a study of costs of regulation. *Journal of Law and Economics,* 5, Oct. 1962

GRAY, F. *The great Canadian oil patch.* Toronto: Maclean-Hunter, 1970

HADLEY, G. *Linear programming,* Reading, Mass.: Addison-Wesley, 1962

— *Non-linear and dynamic programming,* Reading, Mass.: Addison-Wesley, 1964

HAMILTON, R.E. Canadian natural gas export policy. York University, 1971, mimeo

HANSEN, E.J. *Dynamic decade.* Toronto: McClelland and Stewart, 1958

HENDERSON, J.M. *The efficiency of the coal industry, an application of linear programming.* Cambridge, Mass.: Harvard University Press, 1958

HIRSH, W.M. and G.B. DANTZIG. The fixed charge problem. RM-1383. The RAND Corporation, 1954

KENDRICK, D.A. *Programming investment in the process industries, an approach to sectoral planning.* Cambridge, Mass.: MIT Press, 1967

KILBOURN, W. *Pipeline.* Toronto: Clarke, Irwin, 1970

KOOPMANS, T.C. and S. REITER. A model of transportation. In T.C. Koopmans, ed., *Activity analysis of production and allocation, proceedings of a conference.* New York: John Wiley and Sons for the Cowles Commission, 1951

LAXER, J. *The politics of the continental resources deal.* Toronto: New Press, 1970

LEESTON, A.M., ED. *The dynamic natural gas industry.* Norman: University of Oklahoma Press, 1963

LOVEJOY, W.F. and P.T. HOMAN. *Economic aspects of oil conservation regulation*. Baltimore: Johns Hopkins Press for Resources for the Future Inc., 1967

MACAVOY, P.W. *Price formation in natural gas fields*. New Haven: Yale University Press, 1962

— The regulation induced shortage of natural gas. *Journal of Law and Economics*, 14, 1, April 1971, 167–200

NASH, G.D. *United States oil policy, 1890–1964*. Pittsburgh, Pa.: University of Pittsburgh Press, 1968

PLOTNICK, A.P. *Petroleum, Canadian markets and United States foreign trade policy*. Seattle: University of Washington Press, 1964

SCOTT, A. *Natural resources and the economics of conservation*. Toronto: University of Toronto Press, 1955

STIGLER, G.J. The theory of economic regulation. *Bell Journal of Economics and Management Science*, 2, 1, Spring 1971, 4–21

TULLOCK, G. A simple algebraic logrolling model. *American Economic Review*, 60, 3, June 1970, 419–26

WALRAS, L., *Elements of pure economics*. Homewood, Ill.: Richard Irwin, 1965

WAVERMAN, L. Review of W. Kilbourn, *Pipeline*. *Canadian Historical Review*, LII, 3, 1971, 323–5

— National policy and natural gas: the costs of a border. *Canadian Journal of Economics*, V, 3, Aug. 1972

— The dual in linear programming and the preventive tariff. *American Economic Review*, Sept. 1972

PERIODICALS

AMERICAN GAS ASSOCIATION, DEPARTMENT OF STATISTICS. *Gas Facts*. New York: 1967, 1968

AMERICAN GAS ASSOCIATION. *Reserves of crude oil, natural gas liquids, and natural gas in the United States and Canada as of December 31, 1967*. New York: 22, May 1968

CANADIAN GAS ASSOCIATION. *Canadian Gas Facts 1967*. Calgary

CANADIAN PETROLEUM ASSOCIATION. *1966 Statistical Yearbook*. Calgary

DOMINION BUREAU OF STATISTICS. *Crude Petroleum and Natural Gas Production*. Ottawa: Jan.-Dec. 1965; Jan.-Dec., 1966

— *Gas Utilities (Transport and Distribution Systems)*. Ottawa: 1966

FEDERAL POWER COMMISSION. *Annual Reports*. Washington, DC: 1964–8

Montreal Gazette. Aug. 27, 1966

NATIONAL ENERGY BOARD. *Report*. Ottawa: 1965, 1966

— *Energy, Supply and Demand Forecast, 1965–1985*. Ottawa: 1967

Oil and Gas Journal. Tulsa, Oklahoma: 55, 49, Dec. 9, 1957; 57, 30, July 27, 1959; 65, 17, April 31, 1967; 65, 28, July 17, 1967; 66, 15, April 8, 1968

TRANS-CANADA PIPELINES. *Annual Report*. 1957–68.

— *Application October 1967* to National Energy Board

UNITED STATES DEPARTMENT OF INTERIOR, BUREAU OF MINES. *Natural Gas Consumption and Production*. Washington. 1965, 1966

MISCELLANEOUS

CANADA, DEPARTMENT OF TRADE AND COMMERCE. Press release 37/55. Ottawa: Nov. 21, 1955

HOWE, C.D. Speech to the Canadian Institute of Mining and Metallurgy. Department of Trade and Commerce: 1955, mimeo

KERR, J.W. Speech to the Institute of Gas Engineers. Trans-Canada Pipelines Ltd.: May 1965, mimeo

ROYAL COMMISSION ON ENERGY. *First Report*. Oct. 1958

TRANS-CANADA PIPELINES. *Prospectus*. April 19, 1968

— Application to the National Energy Board. Oct. 1957

DATE DUE

DEC 4 '78			
MAY 0 8 1991			

30 505 JOSTEN'S